P9-AOB-871

Families Caring Across Borders

Also by Loretta Baldassar

VISITS HOME: Migration Experiences between Italy and Australia
FROM PAESANI TO GLOBAL ITALIANS: Veneto Migrants in Australia (*with Ros Pesman*)

Also by Cora Vellekoop Baldock

VOLUNTEERS IN WELFARE
WOMEN, SOCIAL WELFARE AND THE STATE (*co-editor with Bettina Cass*)
AUSTRALIA AND SOCIAL CHANGE THEORY
SOCIOLOGY IN AUSTRALIA AND NEW ZEALAND (*with Jim Lally*)
VOCATIONAL CHOICE AND OPPORTUNITY

Also by Raelene Wilding

A CHANGING PEOPLE : Diverse Contributions to the State of Western Australia (*co-editor with Farida Tribury*)

Families Caring Across Borders

Migration, Ageing and Transnational Caregiving

Loretta Baldassar
Associate Professor in Anthropology and Sociology, School of Social and Cultural Studies, University of Western Australia

Cora Vellekoop Baldock
Emeritus Professor in Sociology, Murdoch University, Perth, Western Australia

and

Raelene Wilding
Lecturer in Anthropology and Sociology, School of Social Studies, University of Western Australia

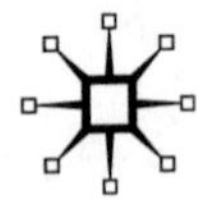

First published 2007 by
PALGRAVE MACMILLAN
Houndmills, Basingstoke, Hampshire RG21 6XS and
175 Fifth Avenue, New York, N.Y. 10010
Companies and representatives throughout the world

PALGRAVE MACMILLAN is the global academic imprint of the Palgrave Macmillan division of St. Martin's Press, LLC and of Palgrave Macmillan Ltd. Macmillan® is a registered trademark in the United States, United Kingdom and other countries. Palgrave is a registered trademark in the European Union and other countries.

ISBN-13: 978-1-4039-4776-5 hardback
ISBN-10: 1-4039-4776-7 hardback

This book is printed on paper suitable for recycling and made from fully managed and sustained forest sources.

A catalogue record for this book is available from the British Library.

Library of Congress Cataloging-in-Publication Data

Baldassar, Loretta, 1965-
Families caring across borders : migration, ageing, and transnational caregiving/ Loretta Baldassar, Cora Vellekoop Baldock, Raelene Wilding.
p. cm.
Includes bibliographical references and index.
ISBN-13: 978-1-4039-4776-5 (cloth)
ISBN-10: 1-4039-4776-7 (cloth)
1. Aging parents—Care—Cross-cultural studies. 2. Caregivers—Family relationships—Cross-cultural studies. 3. Parent and adult child—Cross-cultural studies. I. Baldock, Cora V. (Cora Vellekoop), 1935- II. Wilding, Raelene. III. Title.

HQ1063.6.B35 2007
306.874084'6—dc22

2006047810

10 9 8 7 6 5 4 3 2 1
16 15 14 13 12 11 10 09 08 07

Printed and bound in Great Britain by
Antony Rowe Ltd, Chippenham and Eastbourne

To our transnational families, those who moved away and those who stayed.

Contents

List of Tables

List of Figures

Acknowledgements

Research in the social sciences is always a collaborative effort involving extensive networks. Given the large scale of this particular project, it is not surprising that we have many hundreds of people to thank for assisting with this endeavour. Indeed, we have been lucky enough to have been cared for and cared about across seven countries for over five years as we have brought this book to its conclusion.

Our thanks go to the Australian Research Council for recognising the value of this work and providing the funds that allowed it to be completed. We owe a special acknowledgement to the research assistants who contributed their time, expertise and intellectual energy. Anita Quigley did an excellent job of identifying and developing rapport with the Singaporean families, who, through her professionalism and friendship, soon overcame their reticence to being tape-recorded. Zahra Kamalkhani very kindly allowed us to benefit from her extensive experience with Afghan and Iraqi refugee families living in Perth, and presented fascinating accounts of how these particular groups negotiated their myriad transnational family commitments from both Iran and Western Australia. Stephen Bennetts applied his typical flair and enthusiasm to preparing richly textured interviews, transcripts and field notes of Italians living in Perth and southern Italy. We are grateful for the resulting data, which could not have been gathered without the language, cultural and research skills of these three people.

In addition, we have benefited from the efficiency and attention to detail of those who helped us to transcribe and translate the interviews: Coralie Faulkner transcribed many of the English language interviews with impressive speed and accuracy, Sylvia Kuyer provided excellent and accurate transcripts of the Dutch interviews and the Italian interviews were prepared thanks to the thoughtful contributions of Luisa Biondo, Antonella and Stefania dei Giudici, Vanessa Evangelista, Alessandra Granita, Olivia Mair and Rebecca Sturniolo. The subsequent and time-consuming task of coding reams of data according to the categories we had constructed could not have been done as quickly or as effectively without the assistance of Jan Anderson and Cath Pattenden.

We would also like to thank the many colleagues who have listened to our ideas and contributed their insights. We particularly wish to acknowledge contributors to the Europeans Seminar at the Institute of

Advanced Studies, UWA, and the Moving and Caring Seminar at CAVA, Leeds University in 2003, the World Congress of Sociology in Brisbane in 2004, and the Australasian Centre for Italian Studies Conference in Treviso in 2005, including, in particular, Louise Ackers, Nina Bivona, Ann Baker Cottrell, Anne-Marie Fortier, Donna Gabaccia, Ghassan Hage, Michael Herzfeld, Susanna Iuliano, Khalid Khoser, Cheryl Lange, Jennifer Mason, Isabella Paoletti, Mary Holmes, Zlatko Skrbis, Pnina Werbner, Terri-ann White and especially Ralph Grillo. We are also grateful to the University of Western Australia and Murdoch University for providing the environments in which the research could be completed. Many people at these universities attended workshops and seminars about our research, and we wish to thank them all for their insights and comments, and the experiences they shared. In particular, the members of the Western Australian Migration Research Network (MRN), the department of Anthropology and Sociology at UWA and the Network for Critical Ageing Studies (NCAs) at Murdoch University have been vibrant and responsive groups in helping us to work through our concepts and analysis. The administrative and accounting support of Jill Woodman and Trudy McGlade at UWA have been invaluable as have the efforts of the publishing team at Palgrave. We particularly wish to thank them for recognising the value of this book and for making allowances for the life events, such as new babies, ageing parents and employment obligations, that inevitably tend to affect major writing projects.

Our own families and friends also gave us the essential emotional support as well as the much-needed practical assistance with such things as accommodation, travel arrangements and communicating 'back home'. Special thanks to Ian Wilding for driving Raelene around the winding roads of Ireland and cliffs of New Zealand to interviews (as well as for his enthusiasm for participating in them when invited and for his patience when he was not); to Karen Baldock and Bep Vellekoop (another patient and constant companion/driver) for their wholehearted involvement in Cora's field trip to the Netherlands; to Zita Baldassar, the Bottega and Favero families, Paola Pradal and Domenico Zanchettin for looking after Loretta and her family in Italy. Thanks also to Betty and the Baldassar clan, Lee and Margie Baldock, Brendan, Xavier, Felix, Jennifer and the Jansen family, Victoria Rogers, the Smith family, Kathryn Abbott, Narelle Bartlett and Kelda Mills for ensuring the authors had the necessary support, space, time (and child care) to devote to this book. We are thankful for these and the other many instances of kinwork and friendwork that made our travel and research so enjoyable.

Finally, our greatest debt of gratitude is to the people whose stories are written in these pages. Members of about a hundred families have taken the time to share their experiences of transnational caregiving with us, patiently explaining the joys and the heart-aches, the practicalities and the emotions, of caring for kin who live overseas. Without their willing support and interest, this book would not have been written.

1
Introduction: Transnational Caregiving

> *The best care I can give is to visit as often as I can.*
>
> (Italian daughter)

> *I don't think that my parents will expect me or my wife to support them when they retire, but they will appreciate it if we give them money, you know, filial piety.*
>
> (Singaporean son)

> *My mother needs me more emotionally now she gets older. She would want me closer now, so she can see me more often.*
>
> (Dutch daughter)

> *They [my husband's parents] often say they would love to see more of us, but certainly not in terms of what we could do to help them or anything like that. They probably think that, but they certainly don't express it in that way.*
>
> (New Zealand daughter-in-law)

> *My mother wants to live with me, as I am her eldest son.*
>
> (Afghan son)

> *I don't feel obliged to ring. I just want to do it, and if there's any focus that I have, it's to brighten their day. They're my mum and dad, so I want to talk to them.*
>
> (Irish son)

The above comments are typical examples of concern about ageing parents that were expressed by the women and men we interviewed. They reflect a sense of responsibility and commitment that many people the world over feel towards their parents as they age. There is in fact a considerable amount of research on aged care and intergenerational relations with conclusions that are similar to ours. For example, we agree with other researchers that intergenerational care does not simply go in one direction. Rather, parents provide their adult children with extensive financial, emotional and other support across the life course, and adult children usually begin to reciprocate as their parents get older.[1] Our study also supports previous findings that while both women and men care for and support their ageing parents, there are considerable differences in the extent and type of care that women and men provide (see, for example, Doty, 1995; Joseph & Hallman, 1998). The usual argument is that caregiving is primarily the responsibility of women.[2] Because we interviewed people from a range of cultural backgrounds, we can begin to reassess this claim by acknowledging the obligations that are placed on males to provide care, particularly (but not solely) in non-Western families.[3] The statements about parental expectations of 'filial piety'[4] from the Singaporean and Afghan sons cited above are just two examples.

Our research thus contributes an important comparative perspective on intergenerational relations. But there is something especially remarkable about the 100 or so families we interviewed for this study. Unlike the overwhelming majority of other researchers of intergenerational relations, we are not dealing with parents and adult children who live in close geographical proximity to each other. In fact, these parents and children do not even live in the same country. Our interviewees are transnational migrants living in Perth, Western Australia and their parents 'back home', thousands of miles away, in Italy, Ireland, the Netherlands, Singapore, New Zealand and Iran. Perth is in fact the ideal location for our study, as it is the most isolated capital city in the world, a 4–5 hour flight from its closest neighbouring capitals, including Singapore and Sydney.[5] The vast majority of Western Australians live in Perth, including a very high percentage of residents who were born outside Australia. Since the 1970s, Western Australia's share of permanent overseas arrivals has been fairly stable at 13%, and at the 2001 census it had a higher proportion of overseas-born residents than any other state – more than one-third of its population (Appleyard & Baldassar, 2004, p. 9).

In order to maintain contact and give mutual support to each other, these family members must overcome both the 'tyranny of distance' (Blainey, 1966) and the barriers of borders. In this book, we discuss the ways in which adult children and their parents experience and negotiate aged care obligations and expectations *transnationally*.

Caring across distance and borders

Many researchers argue that only people who live in close proximity – *local carers* – can provide care and support to each other (for example, Rossi & Rossi, 1990; Lin & Rogerson, 1995; Joseph & Hallman, 1998). As a result, very little research has been done on the relationships between ageing parents and their adult children who live at a distance. This is notwithstanding the fact that, especially in large countries with a high rate of geographic mobility, many adult children do not live close to their parents. For example, Moss and Moss (1992, p. 261) have estimated that in the USA nearly half of parents have at least one child living at a distance of 150 miles or more. Studies that have addressed this issue have found that people who live in the same country but at a distance, *translocal carers* as we call them, may make an important contribution to caring for kin who do not live close by (Climo, 1992; Bengtson et al., 1994; Cicirelli, 1995).

Rather than local or even translocal carers, our study is concerned with *transnational carers*, people who live across and care across national borders. Like all caregiving, transnational caregiving is not always easy; the capacity and obligation to give care transnationally is subject to the constraints of time, stage in the family life cycle, health, competing care and work obligations, as well as cultural preferences and expectations. Yet transnational caregiving is also influenced by other factors that make it unique. Contrary to what might seem obvious, while distance is an important factor that often differentiates local, translocal and transnational carers, it is not always critical. For example, in large countries like the USA or Australia, translocal carers might be many miles apart, whereas transnational carers in places like Europe could be living quite close-by, on either side of a shared border, like Switzerland and Italy. Given the isolation of Perth, our sample is characterised by transnational carers who are separated by thousands of miles. These vast distances undoubtedly have important effects on how people care for their kin elsewhere. But just as significant as distance, if not more so, are the national borders across which care must be negotiated, encompassing the realities and regulations of two (or more) nation-states. Transnational carers are faced with visa restrictions and immigration regulations, disparities in the availability of telecommunication infrastructures and of reciprocal health care services between countries, as well as often more expensive travel and communication costs and complicated consequences for employment.

The quotes from transnational migrants at the beginning of this chapter show that they have not stopped giving care to their parents just because they live far away and in different countries: they communicate and visit, send financial assistance and provide emotional support. Furthermore, as the following quotes demonstrate, their parents, where possible, do the same.

> *We sent him money ... the thing is to get them started isn't it? You can't buy a house unless you've things to start with so we've assisted ... with a substantial amount, yes, to help them.*
>
> (New Zealand mother)

> *They're adding to the house, so I'm going to go over and help with that. I love it, I don't like sitting idle so it will be great getting out there and keeping my hands busy.*
>
> (Irish father)

> *Although my mother is not very comfortable herself, whenever I call she wants to make sure we have a warm house and enough food. She is very concerned about us.*
>
> (Afghan son, talking about his mother in Iran)

> *If you have money, you send them overseas to study. So every year we used to send her a lump sum of money, including fees, lodging, and so on.*
>
> (Singaporean mother)

> *I wish my son had not migrated ... But, ... at times there is also a need to give help by hiding your feelings. When things are hard for him, I try to be encouraging. I say, "You've made a choice, you have to be strong."*
>
> (Italian mother)

> *I do it for her. I am the dad. You know, on the answering machine, "Dad, could you do this and that for me?" That's what I'm here for. And I do that with pleasure. For me it is, yes, also fun.*
>
> (Dutch father)

Some types of care are more easily provided by transnational carers than others. Frequent face-to-face contact and regular hands-on assistance are clearly not possible between family members who live far apart from each other. There are, however, other means of care or 'mutual support'. Litwak and Kulis (1987, p. 659), measure them as follows:

> (a) frequency of telephone contacts, (b) frequency of services that do not require face-to-face contact, (e.g., advice and emotional succour), and (c) frequency of services that require only limited face-to-face contact (e.g., help during acute illness, death of a spouse, birth and marriage).

Research that compares the caregiving provided by local and non-local kin shows that local kin are more likely to be aware of the health concerns of ageing parents and provide corresponding 'routine' care. Non-local kin, on the other hand, are more likely to see the cumulative effects of small changes in their ageing parents over long periods of time, and provide 'sporadic' or 'backup' care appropriate to those shifts (Matthews & Rosner, 1988; Moss & Moss, 1992). In this context, Litwak and Kulis (1987, p. 650) have argued that 'geographic proximity is not a simple dichotomy' in which local kin provide care and non-local kin withhold care. Rather, family members take distance into account in deciding how best to take care of other kin. The constraints of national borders are also considered in these negotiations. If caregiving is extended beyond an assumption that it refers to physical 'personal' care activities, such as bathing, cooking and feeding, and instead is understood to include telephone calls, visits, advice, financial support and occasional but intensive help with such things as household repairs and major purchases, then distance and national borders are not necessarily the most important factors in deciding who does the most care (Litwak & Kulis, 1987; Matthews & Rosner, 1988; Joseph & Hallman, 1998; Stafford, 2005).

Our study shows that new communication technologies, such as email, mobile phones and digital cameras, have a very particular role to play in maintaining and enhancing transnational family relations (Wilding, 2006). We examine the way in which the relative availability and affordability of these communication channels, as well as the increased speed, safety and accessibility of air travel, have revolutionised many people's experience of long-distance family contact. We analyse how people rely on these innovations to care for each other, making our study one of very few that document the importance and impact of new technologies on the lives of *transnational* families.

This, then, is a study about *transnational caregiving*; a form of caregiving that can be very special and lead to a quality of interaction not always found between family members who are proximate or living in the same country. Many interviewees in fact made statements such as this:

> *When we lived close together, I had to remind myself. It was a kind of obligation. Now I am more joyful, more relaxed in the relationship. It is a celebration when you see each other. It is a choice. That is the consequence of distance. It creates quality.*
>
> (Dutch daughter)

The aims of our study

The aim of our research is to explore the significance of family members as sources of financial, emotional, moral and practical support across geographical distance and national borders. With a variant on Gubrium's (1997, p. ix) inquiry, we ask how care is 'accomplished' between family members who live far apart and in different countries. We do this through a systematic ethnographic study of the mutual caregiving practices between Australian immigrants from Ireland, Italy, the Netherlands, New Zealand and Singapore and their parents back 'home', and between refugees from Iraq and Afghanistan and their parents living in the transit country Iran. In the study we explore specifically what kind of caregiving and support parents in each of these countries give to their transnational migrant and refugee offspring, and the extent to which transnational migrants and refugees reciprocate such care and support as their parents age. With its focus on the *practices* of transnational caregiving, our approach is similar to that of Ackers and Stalford (2004) who are, to our knowledge, the only other researchers to have undertaken a systematic study of what they call '*doing kinship*' transnationally, in their case between the member countries of the European Union.

While we found the greatest differences in caregiving practices were between migrants and refugees, differences emerged, sometimes quite strongly, sometimes more weakly, between the various migrant and country groups as well. The challenging question of the relationship between national-origin groups and cultural groups is examined in the next chapter, where we consider the methodological implications of the research.

What motivated our inquiry

The inspiration for this research came from a unique combination of capacity, obligation and family history on the part of three transnational social scientists engaged in a long-term collaborative study that began in 1996. At that time Cora Baldock, a sociologist whose academic career had led her in 1964 to migrate from the Netherlands to New Zealand, and from there to the USA and Australia, conducted a modest study of transnational migrants who, like her, experienced anguish about the caregiving needs of ageing parents in their home country. She coined the term '*distant care*' to describe the phenomenon of migrants participating in transnational caregiving (Baldock, 1999, 2000). In writing up her findings, Baldock took issue with literature in family studies and gerontology which had assumed that only family members living in close proximity can provide care. She had no background in migration studies and turned to Loretta Baldassar for advice to make sense of her data as an aspect of the migration process.

Baldassar, an anthropologist and second-generation Italian-Australian migrant, had at that time already conducted many years of research in migration studies, in particular tracing the migration histories and experiences of migrants from the Veneto region in Italy to Australia (Baldassar, 1992, 2001, 2004b). In this context, she had employed the transnational methodology of conducting participant observation and ethnographic life-history interviews with members of the same family in both host and home countries. She highlighted the role of the *return visit* to illuminate the existence of long-time and continuing connections of first-generation and subsequent generations of transnational migrants and their families and communities back home. In theorising the return visit as an integral part of the migration process, Baldassar developed the notion of what she calls '*licence to leave*' to help elucidate the myriad and complex factors that impact on the migration experience, including the return visit (Baldassar, 2001, p. 20). The migration process often results in fractious family histories as the migrants and their family members who remain behind develop different and sometimes conflicting views on the migration experience and on appropriate family roles. Baldassar found that those individuals who have license to leave, that is, whose migrations are supported as appropriate choices by themselves and their families, communities and even nations, are likely to enjoy less fractious transnational family relationships than those individuals whose migration meets with disapproval.

Discussions between Baldassar and Baldock about the intersections of their research led to joint papers on theoretical and policy issues in transnational caregiving (Baldassar, Baldock & Lange, 1999; Baldock, Baldassar & Lange, 1999; Baldassar & Baldock, 2000), and culminated in the large-scale cross-cultural and transnational inquiry that is the subject of this book. Raelene Wilding, a sociologist-anthropologist with a special interest in media and communication technologies, joined the project in 2000. Wilding used this opportunity to further develop her interests in how experiences of love are influenced by and influence the use of media and new communication technologies (Wilding, 2003, 2006). As the partner of a British migrant, Wilding also brought her own set of personal experiences of transnational family relations to the collaboration. The opportunities granted to the research team by the Australian Research Council and their universities to carry out this transnational research have been essential. Equally important, though, have been our sense of obligation to our transnational families and our own life experiences based on family and migration histories.

Policy implications

While our research illuminates important life experiences and has resonated with the experiences of others who face the trials of transnational caregiving, it also has important policy implications. Transnational caregivers have remained largely invisible in policymaking. Yet, as we show in this volume, the intricate and ongoing network of interrelations and obligations between transnational caregivers and their families abroad has a deep impact on their quality of life, both financially and emotionally. Furthermore, the number of transnational migrants and refugees who have such responsibilities and desires to care for ageing parents across national borders is on the rise. In many countries, local carers receive assistance to deal with their caregiving tasks. For example, some receive a carers' pension, some have the right to paid carers' leave, and some receive special consideration from employers in family-friendly industries, or respite care and emotional support from local community organisations. As we show in Chapter 7, translocal carers within national borders receive some support from community organisations (see also Spencer-Cingöz, 1998). Such support does not appear to exist for transnational carers.

At the same time, as we will show in this book, public policies also impose considerable institutional constraints on the ability of migrants to provide transnational care; for example, aged migration, which would allow migrants to provide hands-on care for their ageing parents in Australia, is restricted to those who can afford to pay.

Contemporary researchers who investigate the extent of intergenerational caregiving and family support do so in the context of government policies and practices of 'community care' and 'family care' that are based on the assumption that government expenditures on aged care will rise exponentially as populations age, and that such costs can only be curbed by increased reliance on informal family caregiving (for example, Duff, 2001; Ng, Phillips & Lee, 2002). This assumption is embedded in a view of the elderly as a burden on welfare systems (see Hennessy, 1995; Ackers & Dwyer, 2002; Bittman et al., 2004; Grundy, 2005). Public policies ignore the extent of transnational caregiving between parents and their migrant children, and – possibly unwittingly – impede such transnational care. Such informal care is mainly invisible and thus unacknowledged as an aspect of the national economy. Thus, a central concern of our research into the practices of transnational caregiving is to document relevant public policies and their implications. This is particularly important because both the ageing of the world's population and migration issues are critical points of discussion for contemporary policymakers, yet to date little attention has been paid to the ways in which these issues intersect. Our book addresses this gap in both government policies and the theoretical literature.

Some analytical issues

Transnational caregiving is a new area of research, but it does have a number of clear links to existing research as it lies at the intersection of two major research areas: family studies and migration studies.

Family studies

For many economists and policymakers, caregiving is a source of worry. Not only is caregiving difficult to measure because so much of it is done informally and in the privacy of the home (Bittman et al., 2004; Wolf, 2004), but there is also concern that, particularly in the case of aged care provision, less of it is now being done 'at home', with potentially crippling economic consequences for the state (see, for example, OECD, 1988; Hagemann & Nicoletti, 1989; Duff, 2001; Ng, Phillips & Lee, 2002). Yet this latter assumption, that intergenerational relations of support are in decline and will eventually disappear, is not supported by recent research. Recent work in family studies has given us an appreciation of the strength of intergenerational networks and the degree to which parents and their adult children exchange support and care. For example, Bengtson, who conducted the well-known Intergenerational Linkage Project in the USA (Bengtson et al., 1994; Bengtson, Rosenthal & Burton, 1995), concluded that:

> The typical older person in industrial societies is in close contact with kin, has warm relations with them, and is both a giver and receiver of support and assistance.
>
> (Bengtson & Achenbaum, 1993, p. 17)

Bengtson and Achenbaum (1993, pp. 4–5) termed such reciprocal obligations within the family the '*intergenerational contract*'. The nature of this contract is dynamic and shifts in response to changes in the life cycle (see Nydegger, 1991). Families do continue to exchange support and care across the life course (Hareven & Adams, 1982; Arber & Evandrou, 1993); adult children do care for their ageing parents and, although it is less frequently acknowledged, ageing parents do continue to provide care and support to their adult children and grandchildren, based on what Bengtson and Roberts (1991) have called 'intergenerational solidarity'.

The approach taken in much of this recent literature is a corrective on previous research in several ways. First, it emphasises extended family networks, and rejects theories of the demise of the extended family that had been propagated by Parsonian family sociology with its focus on the close fit between the demands of industrial society and the geographically and socially mobile nuclear family, freed of responsibility for extended kin (Parsons, 1965). In this context, 'new' family studies (for example, Bengtson & Achenbaum, 1993) include a critique of policymakers and journalists who have asserted the existence of an *intergenerational conflict* because of resentment by younger generations about the tax burdens they face having to support the increasing numbers of older people. Second, and especially important, is the attention given in many contemporary family studies to reciprocity in caregiving, as a corrective to research within a more narrowly defined gerontological framework which sees older people as dependents, who are only the *recipients* of care – thereby not acknowledging the degree to which older people continue to provide assistance to their adult children in a genuinely mutual exchange of care and support (see Finch & Mason, 1993; Grundy, 2005).

New family studies provide important contributions to our understanding of family relations. However, they suffer from two major shortcomings. With few exceptions (Bengtson & Achenbaum, 1993; Bengtson et al., 2000; Gupta & Pillai, 2000) they have not explored how families in different national contexts and from different sociocultural backgrounds differ in their practices of mutual family support. Also they have not given consideration to transnational family relations and obligations. The assumption remains that geographic distance has a negative impact on families caring for kin, and some even deny the possibility of long-distance family relations *within* national borders, thus linking 'geography and authenticity' (Gubrium &

Holstein, 1990, p. 45). For example, Rossi and Rossi (1990, p. 422) argue that:

> Across all types of help, geographic distance reduces frequency of social interaction; and hence the opportunity to inform or learn about needs and problems; consequently geographic distance reduces the actual incidence of all type of help flows between the generations.

Those scholars who have challenged the notion that geographical proximity is a precondition for family mutual support have also limited their research to intra-national family relations and to translocal carers. The basic premise of our research is that the broad assumptions of 'new' family studies must be extended beyond national borders; this leads us to migration studies.

Migration studies

In the past, most studies of migration have fallen into two main streams; one examining migrants as labourers focusing on the political and economic contexts of their migrations, the other exploring migrants as ethnic populations in specific host-nation contexts with a focus on issues of identity and integration. In both contexts, the word 'immigration' evoked images of rupture, uprooting, and loss of homeland, and an understanding of the migrant's home and host societies as discrete, homogenous entities. This seemingly unproblematic division of space resulted in a tendency to privilege the notion of assimilation as the end-result of migration and an 'either/or' approach to home and host allegiances (Wimmer & Glick Schiller, 2003). More recently, so called transnational migration studies have challenged this premise of discontinuity, recognising that migrants in fact live their lives across borders, and develop and maintain ties to two (or more) homes, even when their countries of origin and settlement are geographically distant (see for example, Bottomley, 1992; Rouse, 1991; Schiller & Fouron, 2001; Olwig, 2002)

While the concept of the 'transnational' first emerged as a way of describing commercial and other organisations whose activities spanned national boundaries, in the early 1990s anthropologists of migration began to use the term to refer to other sorts of practices that they could see also crossed national borders. The notion of a 'transnational' organisation or corporation became replaced by an idea of 'transnationalism', which was defined as 'a social process in which migrants establish social fields that cross geographic, cultural, and political borders' (Glick Schiller, Basch & Blanc-Szanton, 1992, p. ix). This definition emerged out of a growing awareness of the inadequacy of the assumption that migration is a simple movement from one place to another.

In being attuned to the social relations of transnational migrants, rather than simply to the economic factors in their migration, recent migration studies have thus vastly enriched our understanding. This new attention to the ways in which the social fields of migrants encompass two nations – the 'home' and 'host' nations – has most commonly been discussed in terms of migrant and second-generation commitments to ethnic and national identities. Rather than a singular national or ethnic identity, migrants are now acknowledged as having 'dual' or 'hybrid' identities.[6] For some, this might mean never feeling at home, because 'the centre finds itself wherever the migrant is not' (Mandel, 1990; Baldassar, 2001, p. 6).

The transnational migration perspective focuses on mobility and movement, as well as on settlement, and so acknowledges the ongoing transformative impact of migrancy not only on migrants but also on kin who do not migrate and on the communities in which migrants reside (Ostergaard-Nielsen, 2003). This said, most of the relevant migration literature has tended to focus on the public arena of politics or citizenship, resulting in a call for a greater emphasis on 'transnationalism from below'. In addition, those studies that do consider the more quotidian constructions of transnational social fields (for example, Glick Schiller, Basch & Blanc-Szanton, 1992) tend to emphasise individual (usually migrant) and community (usually ethnic) practices and identities. Few studies[7] explore these processes at the level of families, households and the domestic sphere, leading Gardner and Grillo (2002, p. 179) to argue that 'Despite wide-ranging research on transnational migration and diasporas, many aspects have been accorded less consideration than they deserve', in particular what they call 'the transnational domestic sphere'; 'To understand the meanings and implications of transnationalism for ordinary people ... we need also to consider activities and relationships within households and families' (Gardner, 2002, p. 191). Similarly, Olwig (2002, p. 216), in her work on notions of 'home', argues that if

> migration research focuses primarily on the more abstract, symbolic notions of home, as they may be displayed in ethnic organizations and diasporic cultural expressions, it will leave unexamined the

practices of home, as a household or domestic unit, in which many migrants also engage.[8]

Our book highlights this relatively under-developed set of social relations: the transnational interactions in which migrants and their kin abroad participate – social relations that we argue create *transnational families.*

Transnational family studies

The preoccupation and assumption in the gerontology and family studies literature that caregiving requires proximity within a national boundary is primarily the result of the reification of the notion of family as a private geographic domain represented by a household and of the notion of family as a microcosm and primary reproducer of the patriarchal nation state. This approach to families mirrors the way nations and communities were, until the advent of transmigration studies, largely theorised as inherently connected to a geographical place. In contrast, the study of transnational families and caregiving is premised on an ideational concept of family and kin relations (Gubrium & Holstein, 1990). Much like ethnic, national and diasporic identities and relationships, family identities and kin relations can be maintained across time and distance and are not determined by particular localities or by state borders. As Bryceson and Vuorela (2002, p. 10) argue in their definition of the transnational family:

> Families, ethnicities and nations can be seen as imagined communities. One may be born into a family and a nation, but the sense of membership can be a matter of choice and negotiation. One can alter one's nationality and citizenship just as one can alter one's family and its membership in everyday practice. The inclusion of dispersed members within the family is confirmed and renewed through various exchanges and points of contact.

Our study provides fine-grained analyses of the ways in which families and households interact across vast distances and national borders. The resulting idea of the 'transnational family' is intended to capture the growing awareness that members of families retain their sense of collectivity and kinship in spite of being spread across multiple nations. At the same time, it is important not to underestimate the impact of distance and borders on these relationships and on the practices of transnational caregiving.

The notion of the transnational family leads us to highlight another crucial analytical point. Previously, the migrant was considered the main focus of attention in transnational studies. However, our research suggests that there is another set of transnationals who are easily but mistakenly ignored: the people who never migrate, but whose efforts to maintain contact with their migrant kin mean that they, too, become transnationals – perhaps identifiable as 'non-migrant' or 'local' transnationals. We maintain that non-migrants who stay behind in the home country and who have kin across national borders also think and act in transnational ways. While these 'stay-behind' family members are implicated in notions of transnationalism, their identities and activities are rarely analysed in detail. Transnational caregiving is a practice of reciprocal transnational family networks in which all members of transnational families are involved. In other words, parents and other relatives of transnational migrants are transnationals too, and must be given a voice in research on caregiving across national borders. This premise is reflected in our methodology, which we outline in some detail in the following chapter.

In our research, both the migrant transnationals and the non-migrant transnationals have been included as equally important members of transnational families. This is essential for the experiences in our study to make sense, because family caregiving does not and cannot occur in one direction alone. Regardless of whether a family member has migrated or not, they are part of the transnational kin network of (at least potential) mutual family support. Indeed, the neglect of the non-migrant kin strikes us as a remarkable gap in the literature describing transnationalism from below, as important and surprising as the absence of migrant kin in the literature on family caregiving.

We see *transnational caregiving* as a specific form of both 'transnationalism' and 'caregiving'. Bringing contemporary family studies and migration studies together, we argue that transnational migration does not necessarily lead to a demise of intergenerational family relations. In other words, family caregiving and transnational migration are not incompatible. In fact, we postulate that communities of care and support exist across distances and that transnational migrants are members of such communities of care.

Defining transnational caregiving

In this book, we use the term 'transnational caregiving' to refer to the exchange of care and support across distance and national borders. We develop a model of transnational caregiving that builds on the five types of caregiving or 'mutual support' defined by Finch (1989): economic, accommodation, personal, practical and childcare and emotional and moral. In her study of families in Britain, Finch found that all of these types of support are exchanged between all kin members. However, whether support will be given cannot be taken for granted. Finch (1989; see also Finch & Mason, 1991, 1993) discusses this important point with reference to two major concepts, 'normative obligation' and 'negotiated commitment'. Normative obligation to give care is based on notions of duty and responsibility, supposedly at the core of family relations and frequently defined by policymakers as part of the moral fabric of society that everyone should adhere to. Finch, however, argues that the actual support given to family members is not the outcome of what is the 'proper thing to do' morally, but of 'working it out', that is, negotiated commitments (Finch & Mason, 1993, Chapter 3). These are the long histories of relationships and exchange of care and support that cumulatively inform each decision to provide or withhold a certain form of care required by a particular person at a specific time.[9] Throughout this book we make use of the concepts of normative obligation and negotiated commitment, although we draw particular attention to the fact that such norms and negotiations cannot be talked about without consideration for the cultural differences that exist within and between families, communities and nations as to what is seen as the 'proper' thing to do. Further, we introduce a third concept, that of 'capacity' (or 'opportunity').

It is one of the premises of our book that 'transnational caregiving' is qualitatively and quantitatively different not only from 'local' caregiving', but also from 'translocal' caregiving by the very fact that the capacity to exchange care and support takes place across national borders. Specifically, the additional barrier of the nation-state boundary will, we believe, create important differences between transnational care practices and translocal care practices, even when the latter involve significant distances. For example, at the very least, translocal care does not require the use of passports or visas to visit. Transnational care, on the other hand, will always require some attention to bureaucratic limits on mobility.

At the same time, there are some things that we expect transnational caregiving will have in common with both local and translocal caregiving. We expect that all caregiving is given and received according to 'negotiated commitments', within which context the providers and recipients of care understand their relationships to each other and to others. Also, all caregiving is given and received according to specific capacity and cultural obligation to provide care, as a result of micro, meso and macro factors. For example, lack of employment opportunities is a macro factor that could reduce the ability of adult children to give their parents financial support, even if they live locally. At the same time, micro factors such as who controls the finances in a household (see, for example, Pahl, 1989, 1995, 2001a) may limit the opportunities for adult children to provide their parents with financial support even if they are apparently wealthy, regardless of whether they are local, translocal or transnational.

Nonetheless, we argue that the specific circumstances created by national borders have their own effects on the capacities, obligations and negotiated commitments to provide care. In this book, we aim to demonstrate just *how* and to what extent both distance and national borders influence family responses to capacity and obligation and the degree to which distance and national borders shape the practices of transnational caregiving.

Outline of the book

We begin our discussion in the next chapter with an overview of our research methodology and selection of interviewees. In Chapter 3 we introduce the six groups we included in our research. We provide details regarding the differences and similarities we observed among these groups in terms of factors such as reasons for migration; family and community responses to migration decisions; migrants' and refugees' sense of obligation to care for ageing parents; and age care provisions within home or transit countries as they impact on sense of obligation. The subsequent three chapters report our findings, each dealing with a different aspect of transnational caregiving relations. Chapter 4 sets the context: here we apply Finch's model of five types of family exchange to examine the ways in which each group of migrants or refugees and their parents practice transnational caregiving. In Chapter 5 we explore the practices of transnational communication between migrants, refugees and their parents, demonstrating the different ways in which micro and macro factors impinge on such practices. In Chapter 6 we turn our attention to visits as a particular form of communication and caregiving opportunity, examining why, when and how often members of transnational families visit each other in home and host countries. This is followed in Chapter 7 by a detailed exploration of the policies that affect transnational caregiving, with special attention to visa regulations and government policies regarding citizenship. We draw our findings together in Chapter 8, where we develop a model of transnational caregiving and consider the theoretical implications of our work for the study of migration and families.

2
Researching Transnational Caregiving

By its nature, the subject of transnational caregiving demands innovative transnational research methodologies. The experiences and practices we describe in this book are both emotionally intimate and geographically distant, and the need to capture these two experiences simultaneously has meant transforming conventional research tools into different forms. We outline in this chapter the methodologies we have employed to fit this new research context.

The ethnographic interview

Like all family relationships, the lived experiences of transnational caregiving are often complicated and messy. They alter continuously, sometimes remarkably quickly, at other times more subtly across long periods of time. Also, the sense of obligation and capacity to provide transnational care may shift with each new moment of interaction between migrant and ageing parent. The ideal methodology to research such complex and changeable family relations is the anthropological method of ethnography, defined as:

> participating overtly ... in people's daily lives for an extended period of time, watching what happens, listening to what is said, asking questions – in fact, collecting whatever data are available to throw light on the issues that are the focus of the research.
>
> (Hammersley & Atkinson, 1995, p. 1)

It is widely accepted that, ideally, ethnography requires ethnographers to immerse themselves in a single social context for an extended period of time, preferably a minimum of a year, to get a sense of the annual cycle of social life. As a result, most ethnography tends to take place in a particular site, maybe a village, a suburb or an organisation. However, this focus is no longer sufficient to capture the lives of most people across the globe and presents a special challenge to any research that is intentionally transnational (see also Hannerz, 1992; Gupta & Ferguson, 1997). Yet we still consider ethnographies to provide the richest and most useful accounts of everyday life. So, we have had to think carefully about how to adapt the ethnographic method to our own study of transnational family life and transnational caregiving.

For anthropological studies that aim to research geographically fragmented and mobile populations, a common solution is to do 'multi-sited ethnography' (Marcus, 1995; Ortner, 1997; Scheper-Hughes, 2004). The method we adopted for our research is also multi-sited. We conducted interviews with transnational migrants and refugees in Perth, the capital city of Western Australia, and asked them for permission to contact their parents living abroad. We then travelled to each of the parents' countries of residence, where we made contact with and interviewed the parents and other kin. We chose to focus on the parent–child relationship primarily as a way of refining the focus of the study to make it manageable. As will be discussed below, in many instances other family members, particularly the siblings, spouses and children of migrants (the children, in-laws and grandchildren of parents) also contributed to the study. This multi-sited approach was essential, because transnational caregiving is not something that only one person does. Rather, it occurs in the *interactions* between adult children, parents and their kin living in different countries, and it is necessary to understand both sides of the exchange. If we had focused solely on migrants and refugees, we would have risked reproducing the stereotype of the 'passive aged person' and overlooked the transnationalism of people who do not migrate, but remain in the home country. On the other hand, had we focused solely on parents we would not have found out how migrants and refugees negotiate the specific problems associated with transnational caregiving.

In its strictest sense, 'multi-sited ethnography' implies an intensive day-to-day involvement in the lives of the research participants. This is not possible for a project that aims to understand how people in several different countries, and with a range of different migration experiences, practice transnational caregiving. The people in our study do not share a limited physical space, and so we cannot 'sit in the village centre' and watch how they live their lives and exchange mutual support.

Instead, we adopted the ethnographic method in a truncated form that we call the '*ethnographic interview*'. This involves interviewing migrants, refugees and their parents and other kin in their homes and other familiar social spaces (such as pubs and community events), where it is possible to conduct some, albeit limited, naturalistic participant observation. For example, we were able to observe how people organise their homes, including the availability of spare bedrooms for visitors, and where the tools of transnational communication are located – such as telephones, fax machines and computers. We also noted whether and how migrants and their parents display photographs of their distant kin, or indicators of transnational gift giving such as paintings, ornaments or items of clothing originating from either country. Examples include parents' display of tea towels decorated with kangaroos or wildflowers and the infamous Australian 'ugg boot' or sheepskin slippers, either purchased during visits or sent as gifts. Migrants, for their part, display icons and memorabilia of 'home' such as doormats or posters showing tulips (the Dutch), or packets of Bewley's Tea (the Irish), as well as the ever-present photographs, maps and knick-knacks.[1]

In many instances, other local family members and carers would enter the room during an interview and be introduced to us. Sometimes they would stay and take part, offering their own comments and insights. In cases where parents were unable to fully participate due to illness, we interviewed their primary carer, most often a local child. When we were in the home countries, our interviews with parents were often interrupted by phone calls from migrants in Australia, checking if we had arrived. As part of our ethnographic interviews we recorded all of this activity as important in understanding the transnational relationship. But our approach also involved contact beyond the interview itself. We sent some groups of interviewees detailed, formal summaries of our findings. In other cases, we communicated with our research participants about their ongoing experiences of transnational caregiving through emails, telephone calls, letters and Christmas cards. Interviewees themselves often initiated such contacts. We met some people at later, unrelated community events, where we would swap news and information. Some invited us to have a meal or a drink with parents or other kin when they visited in Perth. In all of these cases, we took notes that added to our understanding of the ways in which transnational caregiving is practised and negotiated. In some instances, such ethnographic data gathering, which started with interviews between 2000 and 2003, has continued to this day.

It is important to note that the degree to which our research represents 'ethnography' in the traditional sense varies across the sample groups. For example, Loretta Baldassar's relationship with the Italian families discussed in this research was extensive, based on many years of fieldwork in both Italy and Perth. In contrast, Raelene Wilding visited Ireland and New Zealand for the first time to conduct interviews with parents, spending approximately four weeks in each country. For Cora Baldock, who carried out interviews with the Dutch sample, overseas fieldwork to interview Dutch parents meant a return to her country of birth, where she is a frequent visitor. The same applied to Zahra Kamalkhani, who is of Iranian descent, and Anita Quigley, herself a migrant from Singapore, who joined the research team to interview refugees and Singaporeans, respectively. Finally, Stephen Bennetts, who carried out some of the interviews in Italy, subsequently travelled to Southern Italy to conduct further ethnographic research for a different project.

There are also differences in the interactions between each of the interviewers and their interviewees. For example, Italian parents perceived Baldassar, a researcher in her mid-thirties with young children, as a person with life experiences similar to their migrant children – possibly even a 'friend' of their children. In contrast, Dutch migrants and their parents saw Baldock, an investigator in her mid-sixties who herself has a migrant daughter living in the Netherlands, as a person with experiences comparable to those of transnational parents. Wilding, a researcher in her late twenties, tended to identify more closely with the young Irish and New Zealand migrants she interviewed than with their parents, while Quigley took on the role of a respectful younger relative in her interviews with Singaporean elders, by specifically addressing her interviewees as 'Aunty' or 'Uncle'. Kamalkhani's role as researcher was different again. Given the uniquely difficult circumstances of Iraqi and Afghan refugees living in Iran, their Perth-based relatives had entrusted envelopes with money and other precious gifts to Kamalkhani before she travelled, to be taken to her Iran-based interviewees. In this sense, she acted as a go-between, enabling the remittance of financial and other material support.

Language, too, is a factor that influenced data collection. All interviews with refugees were conducted in their home language. On the other hand, most of the interviews with migrants were conducted in English. The main exception here was the group of Dutch migrants, whose interviewer Baldock is a native Dutch speaker, but several Italian migrants also felt more comfortable conversing in their mother tongue with Baldassar and Bennetts. As a matter of course, all the parent interviews in Ireland and New Zealand were in English, but in Singapore, the Netherlands, Iran and Italy, all parent interviews took place in the local language without the need for translators. Such variability in the context and conduct of interview transactions is, of course, relevant to our findings.

The questions we asked

We interviewed migrants and refugees in Australia, and their parents in homelands or transit countries. Each interview covered the migrants' history and experiences of migration from the perspectives of both migrants or refugees and their parents, and the ways in which migrants and parents maintained their relationship after migration. This included specific questions about quantity and quality of communication through letters, phone calls, emails or faxes, as well as the frequency and quality of visits, both visits by parents to their migrant children's adopted country and return visits to their homeland by migrants. We then asked questions about the type of care that parents and transnational migrants give each other, using Finch's (1989) model of five types of family exchange (financial, practical, emotional, personal and accommodation). Our focus is on the qualitative experience of these care exchanges and we make no systematic attempt to quantify their relative value. Rather, we are interested in when, how and why care is exchanged, which practices are employed and by whom. We also inquired about migrants' citizenship status, their sense of national identity and belonging and about major factors that facilitate or hinder mutual caregiving, such as finances, visa regulations, job requirements, family stage and history and the quality of relationships between family members.

Most migrants and refugees we asked were very keen to participate, because the topics of transnational caregiving and the wellbeing of their parents were close to their hearts and they said they seldom had an opportunity to talk about such things. Some of their friends who heard about the project asked to be included as well. This was especially the case for the Italians, Dutch and refugees. The brief summaries of interview findings we had prepared for particular sample groups were passed along to others who had similar experiences. At the same time, we heard from participants about friends who had decided not to get involved, because they had recently had to deal with the death of a parent overseas and were not yet ready to talk about this. Other nonparticipants were so anguished about their families overseas that they could not talk about them to strangers.

It is important to acknowledge that families rarely practise transnational caregiving to the same extent all the time. It was not uncommon for families to decrease, and even cease, their participation in transnational caregiving for extended periods. One woman anticipated that her migrant son and daughter-in-law would repatriate after the birth of their first child, and was so disappointed when they chose not to that she avoided all contact with them. After some time, however, as is often the case in transnational family relations, she came to terms with their decision and resumed her long-distance caregiving. While our methodology resulted in synchronic 'snap-shots' of transnational family caregiving practice, we were alerted to the ways caregiving practice changes over time, ebbing and waning across the family and migration life cycles.

This said, there are undoubtedly cases of families who do not participate in transnational caregiving at any time. For example, we know of a family that disapproved of their daughter's marriage and consequent migration, and thus refused to visit, support or even accept the names of their grandchildren. In another case, a migrant daughter would not have anything to do with her family overseas because she had experienced abuse as a child. Our methodology limited our ability to include such families, as a different approach would be required to gain systematic access to their experiences. For ethical reasons, given the advanced age and relative health of many of the parents in the study, we decided to gain access to them through their children. In other words, before we could contact the parents, we had to have the permission of their (migrant) children. This has probably resulted in our sample being biased towards well-functioning transnational families. Even so, in considering how people negotiate the moral expectation from kin and communities to provide care across distance, we are able to offer some explanation for the different and changing degrees of caregiving we encountered both within and between sample groups.

Key variables in the choice of sample groups

Our choice of interviewees and, in particular, the countries they migrated from, was designed to enable an investigation of a number of key variables that we expected would impact on transnational caregiving. These include: geographic proximity between host and home country; the immigration eligibility criteria applicable to particular immigrant groups; diverse ethnic, class and linguistic backgrounds; and length of residence of migrants and refugees in Australia. These general criteria resulted in the choice of a sample representing significantly different populations of migrants and refugees entering Australia including migrants from five different countries and refugees from two countries.

Table 2.1 demonstrates how we conceptualise the relationship of each sample group to the key variables in our study. In terms of *length of residence*, the Italians and Irish, and to some extent the Dutch, represent more established migrant groups than do the Humanitarian Refugees from Afghanistan and Iraq. At the same time, it should be noted that our original intention of including an equal number of long-time residents (since the 1970s) and more recent arrivals (since the 1990s) could not be realised. This was due to the fact that we only wanted to include migrants who still had parents alive in the home country. This was often not the case with migrants who had lived in Australia for 30 years or more: their parents were mostly deceased. In terms of *geographic proximity*, the New Zealanders and Singaporeans are clearly much closer to their home country than the Italians, Irish or Dutch. With regard to language, long-distance migrants include two groups whose native languages are non-English, while one of the short-distance migrant groups, the Singaporeans, although mostly fluent English speakers, are also considered Non-English Speaking Background (NESB) in the Australian context. The refugees interviewed, while not geographically very distant from their home countries, were unable to visit easily. In terms of entry

Table 2.1. Key points of comparison for selection of sample groups

	Proximity	**Era of entry**	**Type of visa***
Italy	Far	1950–2001	Assisted Passage Family Reunion (Spouse) Skilled
Ireland	Far	1967–2001	Skilled Family Reunion (Spouse) Assisted Passage No Visa Required
The Netherlands	Far	1950–2001	Skilled Family Reunion (Spouse)
New Zealand	Close	1974–2001	No Visa Required Special Category Visa
Singapore	Close	1968–2001	Family Reunion (Spouse) Skilled Student Visa
Iraqi and Afghani refugees	Medium	1987–2001	Humanitarian Visa (Offshore) Humanitarian Visa (Onshore) Family Reunion (Spouse) Skilled

*Type of visa listed in order of prevalence for each group of interviewees.

visas, the groups between them cover the full range of *migration eligibility categories*, including the special unrestricted category represented by New Zealand.

Migration eligibility criteria

Because of the importance of migration eligibility criteria to the migration experiences of our interviewees, we include at this point some more detailed information on this issue. Australia has a complex and ever-changing immigration history, but immigration policy as it is known today is essentially a phenomenon of post-World War II. Prior to that time, due to the White Australia Policy and the close affinity between the United Kingdom and Australia, the majority of immigrants had been British, with only small numbers from other European countries, and a general rejection of any entries from Asian countries. The British remained majority immigrants until relatively recently, but post-World War II labour shortages in Australia brought waves of non-British migrants, at first primarily so-called Displaced Persons from Eastern Europe, followed by others, including Italians and Dutch (see Wilding & Tilbury, 2004).

Displaced Persons and migrants, who came under Bilateral Arrangements[2] arrived on 'Assisted Passages', were initially housed in holding camps and required to complete two-year government work contracts, before being allowed to find their own work. The Irish migrants in this study who came on Assisted Passages were not subject to such onerous conditions, merely having to fulfil a minimum (two year) period of residency in Australia or repay the costs of their travel. Others (in fact the majority of people arriving in this period), came on so-called 'Unassisted Passages', financed their own journeys, and relied on other migrants from their hometown or region for support during early settlement (Appleyard & Baldassar, 2004, pp. 20–23). The visa categories of 'Assisted' and 'Unassisted' Passages were abolished in 1982; therefore only a small number of our interviewees fit them (mainly Italian and Irish, but also a few Dutch). Currently, Australian immigration policy regarding the entry of foreign nationals who wish to settle in Australia distinguishes between three main visa categories: Family Reunion; Skilled and Business; and Humanitarian.

Family reunion visas

Crock (1998, p. 68), notes that 'until recently, over half of the migrants admitted to Australia each year consisted of family members of Australian citizens or permanent residents'. The number of people entering as part of the *Family Reunion Program* has declined since the late 1990s due to policy changes, but 'family migrants remain the largest component in the immigration program' (Crock, 1998, p. 68). Family Reunion has two primary subcategories: spouse migration (including migration of *de facto* partners) and migration of family members on compassionate grounds (with separate subcategories for aged parents; adoptive and dependent children; orphaned and aged dependent relatives and 'special need relatives'). The entry category of *spouse migration* is the one most relevant to our study: a quarter of our interviewees entered Australia using a spouse visa. Spouses or prospective spouses are usually granted temporary visas to enter Australia, but are then 'required to wait out a probation period of two years before permanent residency is confirmed' (Crock, 1998, p. 69). This waiting period is waived if partners have been married for more than five years or after two years of marriage if they have a child.[3]

Given the nature of our sample (the selection of migrants who have parents living in their country of birth), family reunion involving the immigration of aged parents and other dependent relatives had not occurred in the case of any of our migrant interviewees, although in some instances migrants expressed the desire to bring their parents out to Australia. Most refugees we interviewed had an especially urgent desire to reunite with their parents and other extended kin in Australia. A very small number had been able to sponsor relatives already, while others continued to make application to bring their relatives to Australia under the family reunion program. However, the possibility of aged migration, as a category of family reunion, is extremely limited, mainly due to the unwarranted assumption that the elderly are a burden on the welfare system. We return to this in later chapters, particularly in Chapter 7, where we discuss the policy implications of our research.

Skilled and business visas

The interrelationship between migration and labour politics is one of the most enduring features of Australia's migration policy. As noted by Crock (1998, p. 92) 'Australian governments have played an active role in determining which foreign workers are admitted in the country'. Currently such migrants are admitted under the category of *Skills-based and Business Migration*. This category, like Family Reunion, is subdivided further. It includes 'Skilled Australian Linked' (SAL) migrants and 'Independent' migrants, both subject to a points test that 'gives credit for skills, education, experience, age and fluency in English' (Crock, 1998, p. 93). The points test, which was established in 1979, includes a

'Migration Occupations in Demand List' (MODL) of occupations identified as being in ongoing national shortage in Australia. Migrants whose occupations fall within the list receive additional points. These points are then aggregated across a range of demographic and other categories (including age, marital status, health and language ability) to obtain a total. A minimum is set as a benchmark criterion, and migrants who meet this minimum score are allowed to proceed with their applications.[4] In the case of SAL migrants, points are also given for their relationship with an Australian citizen. Other subcategories relate to people nominated by an Australian employer because of special talents or skills; some of these are only granted temporary visas. Since 1993 some highly qualified refugees have also been admitted under such skilled visa categories.[5]

The various subcategories of skilled migrants mentioned above are all represented among our interviewees, including examples of migrants on temporary contracts. The latter, mainly from the Netherlands, tend to define themselves as 'expatriates'. There is a final generic category of Business and investment-linked migration, which in its current form has only been promoted by the Australian government since the late 1990s. None of the migrants in our various samples fit this category, although it is said to be a very popular one with Singaporeans and other South East Asians.

Humanitarian refugees

Australia is a signatory to the UN 1951 Convention relating to the Status of Refugees, and thus has a long-standing obligation to receive *refugees*. Most who have entered Australia in the humanitarian program (initially called Displaced Persons) have been offshore refugees; they applied for entry into Australia while overseas (often from refugee camps and under various UN programs). These are so-called 'legitimate' or 'authorised' refugees. More problematic in recent times is the issue of how Australia deals with 'onshore' refugees, people who have entered without a valid visa (that is, as 'unauthorised' refugees) and ask for refugee status. Authorised refugees do receive permanent residency, with all of the attendant privileges. Unauthorised asylum seekers, however, are confronted with highly restrictive measures including mandatory detention upon arrival. Since the introduction of the Migration Amendment Act in October 1999, any people who arrive as unauthorised or illegal refugees can at best expect a Temporary Protection Visa (TPV) – even if they are recognised as eligible for genuine refugee status according to the refugee convention. A TPV enables the holder to remain in Australia for a period of 30 months, but if they leave during that time, they are not permitted to return to Australia, as departing automatically revokes the visa. TPV holders have very limited access to government-provided social welfare, education or financial support programmes, and they are heavily reliant upon the services provided by charitable community groups. Further, TPV holders are processed as individuals, and do not have the right to apply for family reunion (see Kamalkhani, 2004, p. 241). All of these factors combine to create bureaucratic barriers and political consequences for transnational family exchange that are unique to the refugee participants in our project.

Our samples of Iraqi and Afghan refugees contain mainly authorised refugees, but also a few people who entered 'illegally' and who, after a long process of assessment, have received TPVs, the special category of refugee visas with restricted rights.

Special category visa

We included one other group of immigrants, people from New Zealand, who do not require visas to enter Australia. They have free entry to Australia under a Trans-Tasman Travel Agreement developed in 1973, and are the only group of foreign nationals in this privileged position. Until 2001, there was an unrestricted availability of welfare benefits for New Zealand residents in Australia, although, interestingly, they had a relatively low level of welfare dependency compared to other overseas-born populations (Khoo, 1994).

The distinction between migrants and refugees

It is clear from our account so far that we make a strong distinction between migrants and refugees, while also acknowledging that each category embraces a number of subcategories. We are aware that there is a valid case for critiquing such a distinction, based on the false assumption that *all* migrants travel voluntarily and have adequate financial means, while *all* refugees are forced to move and are impoverished. Critics of the distinction (see Koser, 1997) argue that not all migrants move by choice or have alternatives available to them, and can be even more impoverished than some refugees.

Even beyond this critique, the term 'refugee' is not an entirely accurate description of the people in the 'refugee' sample of our study. Unlike their parents, who we interviewed in Iran where they live in transit and are not permitted to seek permanent residency or citizenship, not all of the 'refugees' are still fleeing from their homeland or actively seeking permanent safety. Indeed, some are Australian citizens.

Notwithstanding these complexities, we have retained the term 'refugee' throughout this book as distinct from the category of 'migrant'. These terms are intended to identify the different migration paths taken by the people in our study, in order to better enable us to investigate the effects of each migration path on capacities and obligations to care across borders.

The question of nation and culture

In selecting the sample groups for this project, we have used 'nation of birth' as a central category for selecting participants. It is also a concept that we occasionally use to organise our analysis, when we refer to apparent patterns of difference or similarity between, for example, the 'Irish' and the 'Dutch'. However, we are aware that such uses of national identity are problematic for a number of reasons. First, it tends to result in the collapsing of the important analytical distinction between the organisational features of the state, on the one hand, and the ideological processes of nation-building, on the other (Blanc, Basch & Schiller, 1995, p. 685). Second, the use of national identity terms has the unfortunate and erroneous consequence of reifying perceived ethnic or cultural differences, rather than acknowledging their social construction in specific contexts (Eriksen, 1991, p. 127). Third, it can result in 'methodological nationalism', the false assumption that particular cultural traits or processes are 'unitary and organically related to, and fixed within, [geographic] territories' (Wimmer & Schiller, 2002, p. 305). And finally, national categories tend to homogenise and obscure the diversity of gender, class, regional and ethnic background within populations (Dilworth-Anderson, Williams & Gibson, 2002).

In spite of these clear problems with the terminology, we have nevertheless persisted with using these forms for three key reasons. First, national identity might be a contentious category for social scientists and philosophers, but it remains a cogent means of understanding experience for the many people who participated in our project. Thus, for example, people born in the Netherlands would comment on 'what the Dutch do' and people in Singapore would comment on 'how the Chinese feel'. We do not wish to reify these claims into final statements on 'Dutch culture' or 'Chinese culture'; indeed, our samples are too small to do so even if we wanted to. However, we have aimed to capture the emic perspective on national, cultural and ethnic identities in the chapters that follow. We pay attention to how people conceptualise their national, ethnic or cultural identities, and how this informs their sense of obligation to care, or the care practices they choose to engage in.

Second, the Australian state (like all states) also organises the world into national groups, with significant implications for peoples' experiences. In other words, nation-states provide the borders, replete with rules and regulations, which shape the caregiving practices that cross them. For example, a person identified as a New Zealand citizen has a very different experience of entering Australia when compared with a person identified as an Iraqi asylum seeker or a Singaporean business migrant. Finally, by operating transnationally, the lives of the people we interviewed are bounded by at least two nation-states, each with their own set of laws, institutions and practices. Thus, an Italian migrant living in Australia will have very different concerns about their ageing parents in Italy than their Dutch counterpart, because of the different levels of state provision of aged-care services in Italy and the Netherlands, respectively.

Choice of specific interviewees

It was our aim to conduct approximately *thirty* interviews for each sample group, fifteen among migrants or refugees, and another fifteen with parents. We chose migrant and refugee interviewees using a snowball technique, building on referrals made through our existing research networks. Many of these interviews took place with couples, but migrants who had an Australian partner often chose to be interviewed by themselves, without their partner present. Interviewers would ask their interviewees to approach their parent(s) overseas to invite their participation, and pass on the contact details of their parent to their interviewer. Members of the research team then travelled to interview those parents – as couples or singly for those who were widowed or divorced.[6] In some cases, dementia, illness or immobility of the parent made it impossible for them to participate in the research.[7] When that happened, interviewees were asked to provide the contact details of a family member who took or had taken primary responsibility for the care of the migrant's parent; this was usually a sibling of the migrant or of their parent. Thus, in all sample groups, the 'Parent' category includes some relatives who were interviewed instead of parents. Also, as mentioned in our discussion of the ethnographic interview method, in many cases other family members joined the parent interview. This was particularly common among the

Table 2.2. Number of interviews by sample group

	Interviews with migrants in Perth	Interviews with parents in Perth	Interviews with parents abroad	Interviews with other kin abroad	Total inter-views
Italy*	19	4	13	5	41
Ireland	20	1	11	2	34
The Netherlands	19	2	16	3	40
New Zealand	10	0	8	1	19
Singapore	12	0	9	3	24
Refugees	15	0	4	11	30

*This excludes an additional four families drawn from Baldassar's (2001) earlier research, whose stories are included in our findings.

Italians, the Irish and the refugees, and means that the total *number of people* participating in interviews was considerably larger than the total number of interviews.

Whereas we intended to have 30 interviews for each sample group, and 20 for New Zealand, this aim was not always realised (see Table 2.2). In some instances (for example, the Dutch, the Irish and the Italians, we conducted more than 15 interviews with migrants, but could not create a parent match for every migrant, for the reasons outlined above. In other cases, we interviewed migrant couples, who *both* gave permission to interview their parents, adding to the number of parent interviews. These factors all led to some degree of variability in numbers between sample groups. Also, interview groups varied internally in terms of factors such as cultural, class and linguistic backgrounds, age and life cycle stage; these are discussed in more detail for each group in Chapter 3.

The question of gender

The issue of gender is worth highlighting here, as it is particularly relevant to the literature on caregiving and increasingly so to the literature on migration. Overall, we interviewed more female than male migrants; 60 per cent of our migrant interviewees were female. This is due, in part, to the considerable number of female migrants who were married to Australians. As mentioned above, most of these preferred to be interviewed on their own – and often in their mother tongue, which was not always accessible to their partners. The only exception occurred in the case of the New Zealand sample, which contained a disproportionately large number (70 per cent) of male interviewees (see Chapter 3). In the case of parents, we always interviewed more women than men. The large number of widows among the parents we interviewed help to explain this.[8]

The conduct of interviews and data analysis

All of the interviews for this research, including the visits to the home countries, took place between 2000 and 2003. All interviews with migrants and refugees were conducted in Perth, Western Australia. Interviews were taped, transcribed and where necessary translated into English by professionals. Only the humanitarian refugees and their families did not agree to have their interviews taped; instead, the researcher (Zahra Kamalkhani) took extensive handwritten notes during interviews.

In keeping with our ethnographic style, the interviews were semi-structured. We constructed a list of themes that would be covered in each interview, but the extent to which an interviewee expanded on a particular topic was a matter of personal choice. In part because of the lack of recorded interviews, but also because of a preference on the part of the interviewees, most of the interviews with the Humanitarian Refugees took place during multiple visits. This is in contrast to the majority of the migrant groups, for which one interview per household was the norm. The interviews were usually between one and two hours in duration, although occasionally they stretched to four or five hours. Almost all were also accompanied by additional untaped conversation before and after the formal interview.

All of the transcribed interviews were imported into the qualitative software package NUD*IST for coding according to the main interview themes. These nodes were further coded for more refined themes during the analysis process. The variables we mentioned earlier – relative proximity between home and host countries; visa entry categories; cultural, class, gender and family life cycle differences between and within the various sample groups – were used as organising principles for analysis throughout. For the purpose of analysis we introduced one further set of categories to describe the various factors that impact on the capacity and sense of obligation for transnational caregiving. These were: macro-structural factors, related to government and other institutional barriers which impact on the capacity to care across national borders; meso-factors involving community attitudes and support structures that affect both capacity and sense of obligation to care; and micro-factors, related to family history, attitudes of other family members, gender, family life cycle, as they influence family negotiations regarding obligation to provide intergenerational care.

These categorisations are central to our analysis throughout this book. In order to understand the importance of each set of factors to our research participants, it is necessary to provide further information about each of the sample groups. In the next chapter, we introduce and describe in detail the relevant features of the six sample groups that contributed to our research.

3
Contexts of Migration and Aged Care: Research Case Studies

In this chapter, we provide an overview of the six sample groups included in our study, in order to set the scene for the more general comparative discussions in the later chapters. For each of the groups, we review the history of the migration process and describe the relevant general characteristics such as age, gender and marital status. We then give a brief summary of caregiving facilities in the countries where parents reside, together with some comments on the moral and cultural expectations and obligations regarding the care of ageing relatives. We begin with an overview of the three groups of migrants whose home countries are located at the greatest geographic distance from Australia: those who came from Italy, Ireland and the Netherlands. We then discuss the features of the two groups whose home countries are closest to Australia: those from New Zealand and Singapore. Our overview is completed with an account of the unique features of the refugee sample of people from Iraq and Afghanistan.[1]

'The other side of the world': Italy, Ireland and the Netherlands

Italian migrants and their parents

The best care I can give is to visit as often as I can.

(migrant)

I want all my children with me.

(parent)

Italians have been migrating across the globe for centuries. Of the more than 26 million who left the peninsula since Unification in 1861, most followed seasonal migration patterns to neighbouring European countries (Rosoli, 1978; Bosworth, 1996). North and South America provided the next most popular destinations (Gabaccia, 2000). Since the early 1900s, Italians also migrated to Australia in noteworthy numbers, but this country only began to feature in the broader history of the Italian diaspora after World War II. Australia's need for blue-collar labour inspired the massive post-World War II Immigration Settlement Scheme. Facilitated by the 1952 Bi-lateral Accord, Italians formed the bulk of arrivals in the 1950s and 1960s and had the dubious honour of being the first formally accepted of the 'less preferred' non-British and non-Northern European immigrants (Baldassar, 2004b). Since that time, Italians have been the most numerous group from a non-English speaking background in the country (Jupp, 2001). Their arrival marked a turning point in Australian social policies from Anglo-Celtic assimilationism towards a multicultural approach in service delivery, which has become a feature of contemporary Australia.

The dismantling of the 1901 Immigration Restriction Act (colloquially known as the 'White Australia Policy') in the 1970s saw the removal of any official criteria based on notions of race or colour in the immigration program, but coincided with restrictions on other grounds, including a focus on business and professional migrants and a reduction in overall immigration numbers. These changes, together with the economic development of Italy since the 1970s, resulted in a rapid decline in Italian migration. Italy has since become a host country for migrants from Africa, the Middle East, Asia, Eastern Europe and the Balkans, rather than a place of departure for Italians (Bonafazi, 1998; King & Andall, 1999).

The Italian sample

Our Italian sample reflects the above immigration history and can be divided into three main cohorts of migrants (see Table 3.1). The oldest group comprises post-war, peasant-worker migrants who arrived in Australia in the 1950s and 1960s. This group is made up of four couples aged in their 60s and 70s with adult children and young grandchildren living in Australia. The women migrated through the family reunion scheme, sponsored by their husbands, who had in turn been sponsored by kin or townsmen utilising chain migration networks. Migration had initially been seen by this group as the only way to escape poverty. The objective was to raise enough capital while overseas to fund a successful repatriation to Italy (some had attempted a return to Italy but eventually decided to re-migrate). All have now been permanently settled in

Table 3.1. Italian sample

	Male migrant	**Female migrant**	**Migrant couple**	**Other migrant kin**	**Male parent**	**Female parent**	**Parent couple**	**Other kin in Italy**
Recent families	2	4	4	—	1	6	4	2
1970–1980s families	2	7	—	—	2	3	1	3
Post-war families	2	2	—	2	—	2	—	3

Australia for many years, after the vast majority became Australian citizens in the 1960s, foregoing their Italian citizenship.[2]

The second group is slightly younger, comprising blue-collar and semi-skilled migrants (seven women and two men) who migrated with their Italian spouses in the 1970s and early 1980s. They are now in their 50s with adult children living in Australia and parents, all of them ailing, living in Italy. Two have ties to the earlier wave of migrants through kinship and marriage, and their settlement was facilitated by links to the established Italian communities. The remainder migrated through the Skilled Migration Scheme on two-year employment contracts. The varied motives for migration in this group included individual lifestyle preferences (to escape restrictive family, to flee an uncertain political climate, to find adventure) and, importantly, the potential for better work opportunities. Interestingly, while all of the individuals in this group are long settled in Australia, only the two men have opted for Australian citizenship.

The third cohort comprises ten skilled and professional migrants ranging in ages from their 20s to their 40s, who migrated more recently, mostly in the 1980s and 1990s. Their parents include both young–old and old–old whose health ranged from very well to very frail. All of the people in this group have already taken up, or intend to apply for, dual citizenship, a privilege that has been available since 1992 (Baldassar, 2004b). Most of the migrants in this group had come to marry Australians, or travelled alone and subsequently married Australians, while the remainder came for lifestyle and career reasons.

All of the Italian migrants we interviewed live in the Perth metropolitan area, except for two who live in rural areas not far from Perth. The parents come from various regions throughout Italy, including

Lombardy, Veneto, Tuscany, Lazio, Abruzzi, Campania, Calabria and Sicily, with about half from the south and half from the north. Most parents live in small urban centres; only two in large cities and two in more remote rural areas.

Italian migrants and their life experiences

While they have many things in common, each of these three cohorts of participants have unique experiences that set them apart from the others (for a fuller comparison of these cohorts see Baldassar 2007).

Post-war migrants: Long-term settlement in Australia. Most of the post-war Italian migrants arrived with farming backgrounds or experience in the industrial labour market, some having migrated previously to European or American destinations. In Australia they joined the ranks of the working classes and became segregated in industries characterised by relatively high rates of self-employment, including agriculture, construction, food and fishing. Clustered at the unskilled end of the labour market, they were often regarded as an economic threat by Australian workers and their organisations, and many encountered overt prejudice until very recent times (O'Connor, 1996; Gabaccia & Ottanelli, 2001; Baldassar, 2004a). As a result, these immigrants tended to become residentially segregated, living in the same areas to give each other support and protection.

This high degree of social 'closure' (or 'ghettoisation') was reinforced by both Australian and Italian migration policies, which, after 1925, required migrants (aside from those who were wealthy) to have a sponsor. This meant that Italian migrants were nearly always connected by family or village ties, which facilitated the development of a distinct ethnic community identity and pattern of settlement and incorporation. Such community formation has been a source of support for Italians, including in their practice of long-distance care. For example, during a visit to Italy, most individuals take the time to visit the family of their townspeople and friends in Australia and conduct what one migrant called a 'postal run', both delivering and returning with gifts (including money) and, most importantly, information about the health and wellbeing of family members (Baldassar, 2001, p. 36).

Post-war Italian migrants were not so much individuals intent on settlement in a new country, as members of transnational households enacting the tried and tested economic strategy of return migration for the benefit of their extended families. Migration was often a family decision employed to assist not only the migrant and his future family but also family members left behind. These Italian migrants, therefore, had licence to leave, although they also had a moral obligation to return (Baldassar, 2001, p. 323).

The 1970s cohort: The difficulty of fitting in. Unlike the post-war migrants, all migrants who arrived in the 1970s and early 1980s (both men and women) had previously held paid jobs in Italy. They felt no economic imperative to migrate, although migration offered improved employment prospects in most cases. All of the women had followed their husbands somewhat reluctantly, as the era of massive migration to Australia was considered to have ended, the Italian economic situation was improving and their families could not easily comprehend their decision to leave Italy. Of these migrants, only the two who had connections to the post-war group identified with the earlier migrants and their associations. They all work in semi-skilled or white-collar professions. However, unlike the recent arrivals, they tend to be employed servicing the Italian communities, for example, as language teachers, welfare officers or ethno-specific retailers. The women in this group are particularly lonely, as they do not feel they 'fit in' with the earlier migrants, but are also of a different generation and social class to the more recent migrants. As one woman explained:

> *I have got nobody like me here; it's very hard to make friends. The old migrants don't have the same culture as me, they're not interested to go to the art gallery or to discuss some books to read. The new migrants, they are too busy and are not interested to talk to someone old like me.*[3]

The absence of Italian community identification and support on the one hand, and the difficulty of finding acceptance in the broader Australian society, on the other, makes this group of migrants especially vulnerable. Their children, having had more formal education in general than the children of the post-war migrants, are more likely to follow careers outside of Perth, which has left this group without much local family support. Their families in Italy had never approved of their decision to migrate, given that in the 1970s and 1980s migration was no longer considered a necessity, a factor that further exacerbates their feelings of alienation from both Australia and Italy. This historical context means that these individuals, particularly the women, want to visit home as often as possible, and yet they find the visits very difficult emotionally. All of the parents of this group are in poor health. These migrants find themselves having to deal with the grief associated with seeing their parents dying and the guilt of not having been considered 'good' children. One woman actually made the painful decision to cease visiting as she felt her presence upset her dying mother more than her absence.

Recent migrants: By choice and not for money. The most recent arrivals are a mixture of professional and skilled migrants. They differ from the earlier migrants not only in terms of time of arrival but also in occupation and socio-economic status, which has a direct impact on how much money they can invest in distant care and, in particular, how often they can travel to Italy. In contrast to the post-war migrants, they are not connected to each other or other Italo-Australians by chain migration or village-based ties. They tend not to define themselves as migrants but as '*italiani al estero*', 'Italians living abroad', an identification similar to that defined in the migration literature as 'cosmopolitans' (a notion we discuss further at the end of the chapter). Unlike the earlier migrants, they retain *formal* connections to Italy. For example, teachers keep their names in the job-placement system and return to Italy to renew this entitlement when needed, thus ensuring they can return to employment in Italy if desired. They also have access to dual citizenship, and maintain Italian investments, land holdings and business ties. Their friendship groups are not regionally based and they tend not to associate with the post-war migrants, unless they are married to an Italo-Australian whose parents were of this group, although many interact with Australian-born children of the earlier groups. As one explained:

> *I'm not a migrant. Well, maybe I'm a migrant, but not the stereotypical migrant, which is in view here in Australia. Because here, if you are a migrant, you are a peasant. ... You come from [the South], you are short and hairy. This is the stereotype ... I'm a migrant of choice, not of need. I can go back [to Italy]; I still have one foot in the door, because I have my business, I have my house ... I have a car over there. We go back twice a year, so I am not really an immigrant. On the technical point of view, yes, I am an immigrant. I'm an immigrant here in Australia, I work here, I have a passport, I have a house here. I spend 75 per cent of my life here.*[4]

Being in general much younger migrants, they are also more likely to be the recipients of distant care from their parents in Italy rather than the providers of care. Parents who can afford it often provide financial support to assist their children's settlement in Australia (for example, the purchase of a house) and to fund regular visits home. They are also active transnational communicators, making regular phone calls and sending photos and news of family back home. Those who are able are all prepared and willing to visit Australia to assist their children in times of crisis, in particular with the birth of a child, serious illness or divorce. However, these parents, like those of the earlier migrants, also tend to be ever ready to suggest that children return home, no matter how long they have been in Australia. Italian parents of all groups, in general, find it difficult to accept the permanent departure of their children.

Italian parents and their caregiving needs

Italy currently has one of the oldest populations in Europe, with census statistics from the end of the last century showing 16.2 per cent of the population over the age of 65 and 6 per cent over 75 (Trifiletti, 1998). Only a small proportion of these individuals are in institutional care facilities, comprising 1.5 per cent of those over 65 and 3.6 per cent of those over 75. It was not until 1985 that admission into these facilities was opened to the seriously ill, who were previously cared for at home. Even today, the general community perception is that these facilities are for able-bodied and healthy people, primarily widows, who have no family to care for them (Trifiletti, 1998).

Currently in Italy it is mainly family members, and primarily women, who care for aged people, including the estimated 20 per cent who are over 65 and dependent. The most common pattern of informal care provided by the family, made possible by the traditional practice of living in close proximity to kin, is daily visits from daughters or daughters-in-law. This type of informal care is generally considered a normal duty that falls, first and foremost, on adult children. Due to various recent social changes, including an increase in women's labour-force participation, an increase in the number of children not living in their parents' hometown and a decrease in family size, there are a growing number of foreign immigrants employed to take care of dependent elderly on behalf of other family members (Andall, 1999). Institutional care is more expensive and, depending on the region, less available and accessible and generally considered to be less attractive than domestic help. Known colloquially as having *'una donna in casa'* (a woman in the house), this practice preserves the tradition of family care, even as it disrupts the families of the women employed to provide such care.

At the same time, Italy has embarked on a progressive project to improve the provision of aged-care services in all sectors, including institutional care as well as home care (Trifiletti, 1998, p. 185). Notwithstanding this development, for many Italians there is a strong sense of family shame associated with the use of institutional care, which is popularly characterised as 'locking' older people away. Older, more traditionally minded parents are emphatically hostile to such institutions. One woman, clearly mindful of her approaching 'old–old' age, said angrily that people in these institutions become like 'pieces of ice' and that she would not place her husband 'in the hands of other people, only traitors would do this, people who betray their parents' (Baldassar & Pesman, 2005, p. 147). Older Italians' perceptions about their health and wellbeing have been found to be associated with how close they feel they are to their children (MacKinnon & Nelli, 1996). There is a clear tendency in this group of elderly to believe that if they need to procure support services from outside, then their family is not caring for them adequately (Baldassar, Wilding & Baldock, 2006).

While young–old parents had themselves often nursed their parents in their own homes, they were unsure whether their children would do the same for them. A number indicated fatalistically that their children might place them in a nursing home, 'when their time came', although all hoped this would not be the case. One couple explained that they had a son in Australia and two daughters living and working in other regions in Italy; 'we can't expect them to stop their lives and careers and interrupt their families to look after us'.[5] The prospect of moving from their childhood village to live with one of their daughters in a distant city was not an attractive one. However, they conceded that it would be a difficult choice as to whether to go into a home.

In terms of their aged-care needs, the preference for most Italian parents interviewed, regardless of age and health, is to live out their retirement as independently and autonomously as possible, with a focus on participating in the family life of their children and grandchildren. If assisted living is required, most would prefer to be supported by extended 'family care', and the common (and traditional) expectation is that they would move in to live with a child. Any form of 'institutional care' continues to be the least desirable and least available option.

Irish migrants and their parents

> *Of course, these days communication is so wonderful. A hundred years ago you waved goodbye and that was it, you wouldn't see them again.*
>
> (parent)

Like Italy, Ireland is well known as a country of emigration, and has been sending its people around the world for centuries as a strategy for survival. Irish emigrants have tended to settle in Britain and the USA, with Australia perceived as a relatively undesirable, unfamiliar and distant destination that attracted only an estimated 1 in 14 of all Irish-born emigrants (Jupp, 2001; O'Farrell, 2001). Indeed, many of these were sent to Australia as convicts, against their will and to the anguish of family members who were left behind, sometimes forever. While representing a small proportion of the total Irish emigrants, Irish immigrants were so significant in Australia as to be readily acknowledged now as founding contributors of Australian culture and society (MacDonagh, 1996). The relatively recent acceptance and celebration of Irishness, evident in the valorised myth of the Aussie larrikin, the bush ballad and the popularity of the Irish pub, is in stark contrast to the discrimination and hardship that most Irish settlers – particularly Catholics – experienced at the hands of the Anglo majority between 1788 and World War II (Chetkovich, 2004). The numbers of people migrating to Australia from Ireland have been in consistent decline since the post-war period. While the Irish-born represented some 25 per cent of Australia's population in 1871, in the 2001 Census this had declined to 0.3 per cent, although larger numbers claimed Irish ancestry. The vast majority of the Irish-born in Australia are aged 25 and above, with a median age in 2001 of 49 years and a surprisingly large number of Irish-born aged over 65 (21.7 per cent). Since the 1980s, only very small numbers of skilled migrants relocate to Australia from Ireland in search of adventure, sunshine and a better lifestyle.

Irish experiences of migration

The Irish families who participated in this study were very aware of the significance of emigration in Irish history, and both the migrants in Australia and their parents in Ireland tended to talk about their own family's experiences in terms of this larger historical context. While none of the migrants interviewed were part of the pre-World War II migration, many referred to instances of aunts, uncles and cousins who did make earlier journeys to Australia, never to return. One elderly woman, for example, recalled her family's tradition of holding an 'Irish wake' for extended kin who departed for other countries. Such departures were particularly devastating when the migrants were leaving for Australia, which was perceived as the furthest possible distance from Ireland and thus the least desirable destination. This perception evidently continued into the 1960s and 1970s, as indicated by one woman's memory of preparing to migrate to Australia:

> *I had to have inoculations before I went, and my doctor said, 'why, why are you going to Australia?' He said, 'if you go to England a day's pay will get you home, if you go to Canada a week's pay will get you home, but if you go to Australia, you'll never get home. You know, it's the end of the world'.*[6]

One third of the Irish migrants in our study arrived in Australia prior to 1980. The earliest arrival came in 1967 from England, where she had been employed as a nurse. After meeting her husband through work, she followed him to Australia against her mother's wishes. Her mother remained unhappy about this even after she came to accept her new son-in-law as part of the family. Another four interviewees (three women and one man) arrived in Australia in the 1970s, and described using the Assisted Passage scheme as an opportunity to take an affordable 'working holiday' overseas. All four returned to Ireland soon after they had fulfilled the minimum time requirements (two years) of their reduced fare to Australia. However, the sunny climate and employment opportunities provided a strong lure and none remained in Ireland for more than three years before returning to settle in Australia. While family members in Ireland expressed sadness at the permanent departure of their loved ones, they also talked about how much happier these migrants were living in Australia and, in contrast, how miserable they had been in the Irish drizzle during their brief period of return. For one migrant, the limited economic opportunities in her hometown were mentioned as justifying her decision to return to Australia, but economic reasons were not particularly emphasised by any other interviewees.

The remaining two-thirds of migrants had journeyed to Australia in the 1980s and 1990s, and are characteristic of the so-called 'new Irish' (O'Farrell, 2001; Chetkovich, 2002). Unlike the earlier arrivals, the majority have tertiary education – five of the seven women, and one of the three men – and most work in skilled occupations. Their perceptions of and relationships with Ireland are informed by constant communication with family and friends back home by telephone calls, visits and emails. Indeed, the sense of Ireland and Australia being geographically distant is relatively absent for the 'new Irish'. None had stories reflecting a perception of Australia as 'the end of the world', although two women did report parents who struggled to accept the decision of their daughter to move to another country rather than make a life in Ireland.

The fact that both of these women followed husbands to Australia was relevant in ensuring their licence to leave. This was evident in that, when one woman's marriage ended, her family placed her under significant pressure to return to Ireland.

All except two of the post-1980s women migrated to Australia for reasons related to a love relationship, which either began in Ireland or while they were on a working holiday in Australia. Only one man cited economic motives as the primary reason for his migration, and even his reasons were complemented by a desire to partake in the 'Australian lifestyle'. All interviewees talked about their decision to remain in Australia as influenced by such factors as enjoying the outdoors lifestyle over the pub-based social life in Ireland; all were optimistic about employment and housing opportunities, and all particularly mentioned the sunshine. However, another explanation for the decision to remain in Australia is possible for at least some of the post-1980s migrants. It is notable that three of the seven women who arrived after 1980 were either divorced or remarried, and all three talked of how their divorce remained an 'unspeakable' topic and a source of stigma in Ireland. While this was never explicitly stated as a reason for migrating to Australia, they did reflect on the relative freedom with which divorce could be revealed and discussed in Australia, and how this made their everyday lives less fraught.

All but two of the Irish migrants have taken up Australian citizenship. However, none have given up their Irish citizenship to do so. This dual citizenship is also reflected in the multiple identities that are balanced by the Irish-born living in Australia. On the one hand, they report a desire to make the most of opportunities to be Australians, and to enjoy the opportunities that they perceive as being available in Australia in terms of lifestyle and employment. On the other hand, they also report a strong sense of 'being Irish', which includes an acknowledgement of certain obligations to family members who remain in Ireland. While most have parents too young and independent to need care yet, all discussed the anticipation of future requirements for providing care to their ageing parents, and how this might be managed in conjunction with other family members in Ireland.

The sample of Irish migrants and their parents

During a series of 20 interviews, a total of 22 Irish migrants were interviewed in Perth (see Table 3.2). The majority (16) were women, with only 6 men participating. In two cases, interviews were conducted with couples in which both partners had parents living in Ireland. A total of

Table 3.2. Irish sample

	Migrants	**Parents**	**Other kin in Ireland**
Male	4	1	—
Female	14	9	2
Couples	2	2	—

14 interviewees were matched with parent interviews in Ireland; in two cases the migrants did not wish to have their parents contacted for the project, in one case the parent was ill at the planned interview time, and in five cases the migrants were eager to participate in the project but their parents were deceased. These latter interviews have informed the analysis, but direct quotes and case studies have not been incorporated into our discussion in this book.

Almost all of the interviewees were married, with only one divorced without a partner at the time of interview, and another in a *de facto* relationship. Almost all had children, many of them school-aged or younger. The majority of women were employed part time, half of them as nurses and most of the men were employed full time as managers or in skilled trades. Their high levels of education and skills reflect the Irish-born population in Australia more generally, which is above the Australian average at 56 per cent compared to 46 per cent in 2001. All rated their income as middle and upwards. All were resident in the Perth metropolitan area, the majority in homes that they had either bought or were in the process of buying. Although religion was not a question we asked about explicitly, it became clear during these interviews that the majority were Catholic, with two Irish Protestants and two identifying as Anglo Irish.

Historically, particular areas of Ireland have been more strongly associated with emigration than others, due to economic depression. However, such concentrations of regions of departure are not evident in this sample group; this may be a result of the small number of people interviewed or possibly of the shift in motivations for Irish emigration to Australia from economic to lifestyle factors. Parents were resident in rural, semi-urban and urban locations across the length and breadth of Ireland. Four of the parent households were rural residences at some distance from the nearest town; another six were in towns of varying sizes, such as Letterkenny and Monaghan; and the remaining were located in larger cities such as Cork and Dublin. Only three migrants interviewed had both parents alive; most had widowed mothers, one had a widowed father and in one instance both parents were deceased. At the time of interview in 2001, one-third of the parents were 'young–old', in their 60s, one-third were in their 70s and the remaining third were in their 80s. Even among parents in their 80s, mental and physical health was generally of a high standard, reflecting evidence in a 2001 study that almost 70 per cent of older people in Ireland reported their health to be good or very good, while nearly 80 per cent considered themselves to be fully self-sufficient (NCAOP, 2001b). One male migrant reported the difficulties incurred by his mother's onset of dementia, and one woman talked of the challenges of caring for her mother during the final stages of a terminal illness (in these cases, interviews were conducted in Ireland with siblings of the migrant, rather than their parents). However, the remainder of parents are independent and active, and reported only minor, manageable health complaints such as high blood pressure, some loss of sight and hearing and some reduced mobility.

The relative youth and good health of this sample of parents may help to account for their high level of contact with Australia and their significant participation in travel both to Australia and the rest of the world. All of the parents have children living within Ireland, with whom they exchange mutual support on a regular basis. In addition, one-third of the parents have adult children living in countries outside of Australia and Ireland. All but one of these parents have made journeys to visit their children and other relatives around the world. Indeed, some parents reported that having a child in Australia was a great incentive for an extended holiday in a sunny climate. For others, a round-the-world trip once every two or three years had become a pleasurable retirement pastime. However, all parents and migrants commented on the arduous nature of long-distance travel between Ireland and Australia, and agreed that a time would come when the parents would feel unable to undertake such a journey. In fact, some parents had already made their final visit to Australia, and preferred instead to contribute to the fares for their grandchildren and children to visit Ireland.

Irish parents and their caregiving needs

The age distribution of the parents in this study is similar to the ageing of Ireland's population generally. In the Republic of Ireland, 11.2 per cent of the total population in 2001 was aged 65 and over, almost 60 per cent of which were female (NCAOP, 2001a). Not only are the over-65s becoming an increasingly significant proportion of the total Irish population, but those aged 75 and over are becoming a larger proportion of the older population, increasing from 37 per cent in 1986 to a projected 44 per cent in 2006 (Leane, 1995). The ageing of the population, in tandem with recent transformations such as urbanisation and the changing size and composition of households have contributed to discussions in Ireland about what constitute appropriate caregiving arrangements for Ireland's elderly (Larragy, 1993a).

Almost half (48 per cent) of Ireland's older people live in rural areas, a larger proportion than the under-65 age group (42 per cent, NCAOP, 2001a). Yet the relatively recent urbanisation of Ireland has transformed the historical patterns of residence of Ireland's older people. Since the time when the Irish farm was no longer able to support all of a couple's children through their adulthood, Irish traditions of inheritance and residence gave a son, usually the eldest, first preference for inheritance of the family farm and encouraged the remaining children to migrate (Arensberg, 1937). It was taken for granted that the heir and his wife would share the family house with his ageing parents and, in exchange for the inheritance, would provide 'service to the old people' (Scheper-Hughes, 1979, p. 42). While it is now less likely that a newly married couple will share the home of the husband's parents, this preferred pattern of residence and caregiving for elderly people was confirmed in a government policy document in 1988, which recommended that 'as far as was possible elderly parents should be facilitated to retain their independence and dignity within their own homes' (Leane, 1995, p. 59). Most caregiving services provided by local religious, government and non-government organisations are thus aimed at preserving the rights of older people to remain in their own homes with the support of family and the community (Horkan, 1995).

The policy of maintaining elderly people in their own homes relies to a large extent on the assumption that Irish older people will have family members available to either coordinate their care or, preferably, share their home; 'the importance of informal care in Ireland is unmistakable' (Daly, 1998, p. 29). But it is becoming clear that this assumption is not always matched by reality (Daly, 1998; see also Scheper-Hughes, 1979, p. 75). As is the case in Italy, the Irish household size and composition is undergoing change (Kennedy, 1986) with increasing numbers of elderly people living alone (Curry, 1998). There are changing expectations regarding gender roles (Kennedy, 1986), and new ideas about what constitutes a desirable standard of living (Larragy, 1993b). Most Irish people agree that the traditional family farm is no longer able to provide the sort of income required to support a desirable standard of living (see also Scheper-Hughes, 1979). Furthermore, declining birth rates mean that fewer children are available from which to select an inheritor of the family house to live with the ageing parents. Also, more lucrative employment opportunities within the European Union and elsewhere around the world have encouraged entire families of adult children to migrate out of Ireland, leaving many parents without immediate access to their children and grandchildren (Giarchi, 1996). Those adult children who do remain in Ireland often have to relocate to the urban centres, such as Dublin, in order to earn the incomes that provide them with the necessities and luxuries of modern living. Also, women are increasingly expecting to pursue their own career paths and to earn independent incomes; this has an impact on their ability to provide the constant at-home care that was expected in the past.

These social, economic and cultural transformations within Ireland have resulted in the building of more accommodation specifically designed for the needs of older people, including nursing homes, hostels and secure independent accommodation (Larragy, 1993b; Daly, 1998). Older Irish people have also benefited from additional services provided by the government. For example, social welfare support is available in the form of pensions, free telephone rentals, power subsidies, free travel on some public transport and medical cards, among other services, improving the lives of many older people in Ireland (see also, NCAOP, 2001c). At the same time, many ageing parents are concerned about the loss of immediate family support as a result of national demographic and economic changes. For example, one Irish mother in her 80s expressed her sense of relief that she had four daughters, two of whom were living in her immediate neighbourhood. She said that she was the envy of many of her local friends, who had no children living locally and were on waiting lists for the limited places available in newly constructed aged-care accommodation. At the same time, she expressed a sense of frustration at her remote rural location. One advantage of aged accommodation, she suggested, was that they were usually close to urban centres. She yearned for the opportunity to make independent visits into the shops and tearooms of the town without relying on her daughters or resorting to the one bus a day that travelled past her home in and out of town. Other parents in our sample, while not living on working farms, had acquired sufficient landed property to give their children parcels of land on which they could then build their homes. This meant that these parents (mostly widows) were assured of having children in close proximity to care for them as they aged.

Dutch migrants and their parents

After all, your children are only on loan.

(parent)

The large wave of Dutch immigrants who settled in Australia in the 1950s came to forge a better economic future for themselves and their children after the devastation of war and in response to severe over-population during the post-war baby boom. They were working class craftspeople and farmers who had not had the benefit of much education and sought a better life for their – often numerous – offspring. Other Dutch people who migrated in the 1950s arrived directly from Indonesia, preferring Australia to repatriation back to the Netherlands after the loss of the Dutch colonies. Dutch migrants favoured Australia over many other destinations, and were welcomed by Australian authorities as attractive alternatives to migrants from the United Kingdom (Jupp, 2001, pp. 258–260). During the years from 1947 to 1961, over 125,000 Dutch people arrived in Australia, about 30 per cent of all Dutch people emigrating from the Netherlands (only Canada attracting more).

This large influx of Dutch immigrants between 1947 and 1961 was a phenomenon of short duration. In fact, notwithstanding 'the dynamic urge to go out into the world, to acquire and prosper, to explore and understand' deeply ingrained in the Dutch mentality as early as the seventeenth century (Schama, 1987, p. 491), the number of Dutch people who settled abroad permanently was small before the 1950s, and again relatively insignificant in post-1961.[7] According to the 1996 Census, Dutch migrants who arrived in Western Australia since the 1970s make up less than 15 per cent of all Dutch-born residents in the state. These more recent arrivals formed the bulk of the Dutch migrants in our sample.

The Dutch sample

The Dutch migrants in our study were comprised of three categories (see Table 3.3). Firstly, we interviewed women and a few men who migrated to marry Australians. These were generally young people (in their early to mid-20s) who travelled on their own, on visitor or spouse visas, to join their Australian partners. Secondly, we included so-called expatriates, mostly professionals sent out to Australia on limited contracts (usually for four years) to work for major international companies such as Shell. Aged in their late-20s to mid-30s, they tended to arrive as a family group with small children, and expected to move on to other posts as part of the international workforce of large global enterprises. Some of this group, however, wanted to stay in Australia and sought permanent residency status, hoping that the company that had sponsored them would offer them long-term employment in Australia. Finally, we interviewed male immigrants with their immediate family

Table 3.3. Dutch migrant sample

Migrant group	Number of interviews	Female migrants only	Male migrants only	Couple interviews
Brides/grooms	7	5	1	1
Expatriates	4	1	—	3
Settlers	8	4	1	3
Total	19	10	2	7

whose motive to migrate was a career change, coupled with a need to escape from overcrowding, pollution and what they saw as excessive bureaucracy in their home country. The breadwinner in these households generally gained entrance as a skilled migrant, because of advanced skills in areas such as engineering, computer science and other 'marketable' professions. Not many women fitted this category, confirming data from other studies (see Smits, 1999, 2001) that Dutch women seldom migrate for their own career. In fact, several women we interviewed had used their influence to try and prevent or postpone their husband's migration plans. Those few women who migrated in their own right came also for a change of occupation, but in their cases, coupled with a sense of adventure or a need to escape from awkward family relations. Age at arrival varied for migrants in this group, although due to Australia's migration criteria, which give preference to young people, most were under the age of 40.

All Dutch migrants we interviewed live in metropolitan Perth. On the other hand, interviews with their parents in the Netherlands required travel throughout the entire country. One-third of parents lived in small villages, another third in middle-sized towns and the remainder in big cities. It is worth noting that migrants brought up in small communities in the Netherlands had moved to larger towns for study or work before eventually emigrating abroad. Other research has shown that people willing to make such long-distance career moves (even within the Netherlands) are a strongly motivated group with higher than average earning potential (Smits, 2001, p. 558).

As indicated in Table 3.4, about two-thirds of parents were 'young–old'; they ranged in age from late-50s to late-70s and, except for one, were in good health. Their migrant children were mostly young people with small children who had departed in the late 1980s and 1990s. Only four parents were 'old–old', in their late-80s, and suffered from physical ailments and/or dementia. At the time of interview, their

Table 3.4. Dutch parent sample*

Interviewees	Number of interviews	Mother only	Father only	Parent couples
Young–old	12	4	1	7
Old–old	4	3	1	—
Total	16	7	2	7

*Two interviews were conducted with daughters residing in the Netherlands, whose mother had migrated to Australia, and one with a local son whose mother had recently died. For the sake of clarity, these three interviews have been left out from the table.

migrant children were mostly middle-aged people, who had left the Netherlands in the 1960s and 1970s.

The parents we interviewed were all financially relatively well off. This is due in part to the still quite generous provisions for old age current in the Netherlands. Such generous provisions have two important consequences: parents have little or no incentive to move to Australia, and migrants can feel confident that in case of old age problems their parents are well looked after, both financially (through their pensions) and in terms of housing (through a range of sheltered accommodation for the aged). We return to these issues below.

Dutch immigrants and their life experiences

Common to the three groups of Dutch immigrants was their experience of the Netherlands as a cosmopolitan, individualistic culture with a longstanding tradition of transcultural mobility. All Dutch migrants in our study had enjoyed travel to other European countries; they speak English well and, once in Australia, most appeared to become integrated relatively quickly. With few exceptions they did not seek association with members of the Dutch–Australian community (first and second generation) who had settled in the 1950s. Most are well educated and from a professional background. At the same time, the three groups have certain unique features that set them apart from each other, which allow a special insight into the variability of migration processes and differing senses of national identity, in turn leading to variable experiences with transnational family caregiving.

Falling in love with Australians. In the first few years after migration, those Dutch migrants (mainly women) who came with the specific aim of marrying Australians faced unique problems not encountered by others. They came typically on a tourist visa for 6–12 months. Possibly because of the Australian government's concern about sham marriages (Crock, 1998, pp. 68–79), these migrants reported that immigration authorities strongly urged them to get married before these visas expired, as this would strengthen their case for a Temporary Spouse visa. As one woman, who had migrated in the 1970s on a six-month tourist visa, explained:

> *After six months the only way I could stay here was by getting married ... and I didn't really want to get married yet, but ... it was a matter of getting married or leaving the country. I believe I had already made the decision, I wanted to stay, but [my partner] needed another week to think about it, which was very scary for me, because I knew, if he would say no, then I would have to leave.*[8]

Once they received their Temporary Spouse visa, the women in this migration category reported that they were obliged to remain in Australia for a probationary period of two years before being granted permanency, during which time they could not engage in paid work. Not being able to take on any kind of work meant that they felt very lonely and isolated. Not being able to contemplate a return visit during their first two years added to this sense of isolation. Several women described their state of being during these first years in Australia as akin to depression. It is noteworthy that one of the two men who migrated because they fell in love with an Australian woman had deliberately applied for an 'independent residency visa, which was so we did not have any pressure of having to get married'.[9]

The parents of these migrants were still quite young when their daughters (and sons) embarked on their 'romantic journey', and many visited during the period their migrant children were not able to leave Australia.[10] Such visits helped to alleviate parents' concerns about their daughters' wellbeing; some mothers also came to assist with the birth of children. When these migrants began making return visits, they were usually accompanied by their young children (husbands did not always come along). The caregiving needs of parents were not yet an issue. Instead, the focus of visits and return visits was the opportunity for grandparents to become acquainted with their grandchildren. Migrants in this category sometimes expressed guilt about the fact that their decision to migrate deprived their parents of frequent access to their grandchildren. In turn, grandparents felt upset not being able to help with childcare.

This category of immigrants is possibly most affected by the push and pull of two cultures, their sense of identity wavering between the

Australian mainstream and their Dutch heritage. Their partners generally do not speak Dutch, and their children, even if they learnt Dutch when they were little, tend to lose the ability and the will to speak the language as they grow up. This leads to communication problems with grandparents, and the need for the Dutch migrant to negotiate between their parents back home and their English-speaking partner and children. Interestingly, all of these migrants have (so far) retained the Dutch nationality. They would prefer dual citizenship, but this had not yet been available to them.[11]

Moving across the globe. The accounts of the Dutch expatriates we interviewed provide insights relevant to the life stories of any professional people who move across the globe for career reasons. The male expatriates had known from the time they did their university training, or signed up for a particular company, that postings in distant locations would be part of their lives. Their career paths in many ways required continued mobility – sometimes closer, sometimes further from 'home'. Although their partners may not have had the same interest in constant mobility, they knew well enough that being married to a geologist or mining engineer meant they were expected to move about. The father of a male expatriate working as a geologist made this very clear:

> *If you want to have a career, you have to go abroad, with that company. So you just need a woman then who wants to go with you ... and if she doesn't want to go with you, then you can't have a career.*[12]

The expatriates were the most privileged of all migrants in that they earned high incomes and had ample opportunities for regular (industry-paid) return visits. Their parents also appeared to have the resources for frequent visits and they came, as in the case of the migrants who married Australians, to see their grandchildren.[13] Possibly because the expatriates can never be sure where they will reside next, their sense of Dutch identity remains strong and they said that they would retain their Dutch nationality at all costs. The provision of Dutch-language classes to their children, paid for by their employer as part of their contractual arrangements, ensures that their children also retain a close link to their home country.

In interviews, the dependent wives in this group appeared self-assured and matter-of-fact about their distant parents; and their parents back home did not express the sorrow and sadness shown by the parents of migrant brides. Several factors are probably at play here: the wives of expatriates derive a sense of security from their financial status, and their opportunities for frequent visits back home; their parents feel assured that their daughters are moving across the globe with a Dutch partner (rather than settling in a strange country with a 'foreigner'); and there is a general sense of taken-for-grantedness, about transnational mobility and eventual return to the homeland, that makes it easier for parents of expatriates' wives to accept their loss.

Settling down. Those Dutch migrants who settled in Australia with their immediate family mostly continued the traits that Dutch migrants are known for: successful integration and easy acceptance within the Australian community (see also Peters, 2004). Like the expatriates, most moved as a family, usually with children, and had jobs waiting for them when they arrived. Only a few of the dependent wives in this group, who had migrated reluctantly for the sake of their husband's career, initially experienced a sense of isolation and loneliness. They were probably helped by the fact that, unlike the migrant brides, they were not restricted in their options: their migration status allowed them to take up paid work from the start. One of them, according to her mother, 'would not have made it, if she had not had her work'.[14] Those wives who did not take up paid work because of their family commitments soon became active in their children's schools or playgroups.[15] The migrants in this group were most likely to take out Australian citizenship, although there were some exceptions; several women (but not men) preferred to retain their Dutch identity. Although fluent English-speakers, these migrants, like the expatriates, would continue to speak Dutch in the privacy of their own home and among friends. It is important to them that their Dutch-born children will retain their knowledge of the Dutch language, but they are realistic in their expectations. They know that once children attend Australian schools, they will soon forget or refuse to speak Dutch.

When these migrants left the Netherlands, in most cases their parents were still quite young, and able to visit. However, when we interviewed them, in some instances 20 years or more after their arrival in Australia, most of their parents did not visit any more, due to their advanced age. In fact, those we classified as 'old–old' were all parents of settlers, and they were all widowed. Even when parents were still in relatively good health, they had become somewhat fearful of travel. In this group of migrants we encountered the greatest anguish about their parents' health and wellbeing. Some had recently lost one of their parents, and the issues of caregiving, death and dying were very much on their minds.

Dutch parents and their caregiving needs

There are several important features that distinguish care for the elderly in the Netherlands from aged care in the other countries in our study. First, the Netherlands is a highly regulated society, where the government maintains strict control over standards and practices. This means that paid aged care services must be of a high standard, and the government prevails if people cannot afford to pay for institutionalised care in *'verzorgingshuizen'* (hostels) or *'verpleeghuizen'* (nursing homes). Such institutional care has special features. Residents in Dutch hostels have their own separate apartments with their own belongings, and live as if in their own home, while able to access nursing care and cooked meals. Only disabled or chronically ill elderly reside in nursing homes. Such nursing homes tend to be divided into two separate wards: one for physically disabled people (wheelchair or permanently bedridden), the other for psycho-geriatric patients (some of whom may of course also have physical ailments). Most are non-profit institutions provided by various church organisations.

Secondly, there is a strong sense that older people need to retain their independence from their families. This is exemplified by government policies and practices of ensuring that older people remain in their own homes, or in independent dwellings especially designed for seniors. These include: flats for seniors; clusters of homes for people over the age of 55; *'aanleunwoningen'* built adjacent to hostels where older people retain independence but can avail themselves of cooked evening meals; and service flats. Only well-to-do seniors can afford a service flat – all services provided there have to be bought.

Finally, and importantly, the need for independence expressed in Dutch-aged care policy is also felt strongly by older people themselves, and their definition of independence appears significantly different from elderly parents of the other migrant groups (Baldassar, Wilding & Baldock, 2006). Dutch surveys have shown that when given a choice between paid care and informal care by children, other relatives or neighbours (*'mantelzorg'*[16]), the overwhelming proportion of elderly indicate they prefer *paid* professional care. They do not wish to be beholden to their children, and do not wish to live with them, or be dependent on their children's care (Timmermans et al., 1997, p. 173). This means that most people in the Netherlands will not move in with their children when they require care, nor will their children move in with them. They will try to live close to, but not with, their adult children.[17] Movement to a hostel is seen as a positive choice. The design of hostels reinforces this sense of independence: all residents have separate apartments with kitchenettes and *en suite* bathrooms; such apartments have their own front door, with nameplate and letterbox (Bijsterveld, 1996, p. 174). Nursing homes, however, are feared for what indeed they are in the Netherlands – as 'places to die'.

The parents we interviewed in the Netherlands all lived independently, mainly in their own homes or apartments, and in a few cases in special dwellings designed for seniors, including one service flat, one *aanleunwoning* and one unit in a cluster of homes for seniors. One 'old–old' parent, who still lived in the family home, had a stair lift installed and safety rails in the bathroom. In the two cases where parents had recently died, this occurred in hospitals after a period of intensive caregiving at home by their surviving spouse and other relatives. When asked about their needs for care, most parents, regardless of age and level of disability, indicated that they do not want to be a burden on their family, and especially do not wish to upset their migrant children with their problems. None expressed any desire to live with their children. The language that some parents used in expressing their views is interesting. They would say 'after all, your children are only given to you on loan'; therefore, children 'have to go their own way'. As Schama (1987, pp. 481–561) describes with great insight, Dutch parents from the sixteenth and seventeenth century onwards, steeped as they were in the Calvinist faith, expressed themselves in painting and literature to be strongly protective of their children, aware of their vulnerability and deeply grieving whenever any of their children died, but at the same time stoic, and accepting the inevitability of their children's separateness and future independence. It is quite possible that parents' views that 'they only had their children on loan' are in fact remnants of this Calvinist heritage.

It is important to stress that, notwithstanding parents' wish not to be a burden, adult children in the Netherlands do carry considerable responsibilities towards their ageing parents. In fact, older people who are not institutionalised (the overwhelming majority of seniors) rely mainly on their children (as well as neighbours) for assistance with things they cannot readily carry out themselves (see, for example, Knijn & Kremer, 1997). Only approximately 8 per cent of the Dutch elderly live in hostels or nursing homes (Timmermans, 1996, p. 17). Most families are fairly close-knit, and it is common practice for adult children to visit regularly, assist their elderly parent(s) with shopping, accompany them on visits to doctors and take care of household repairs where needed. Their care continues when parents become institutionalised, in that they are often consulted by staff about their parents' needs, continue their regular visits and may even get involved in personal care duties during such visits, although the latter would be a matter of personal choice as paid staff are always on call. The views of the Dutch migrants we interviewed reflect this. They regretted, and felt guilty about, not being able to give hands-on assistance to their parents on a day-to-day basis, and would make every effort through letters, phone calls and visits to provide emotional and practical support to both their parents and any primary caregivers among their siblings back home. It should be added that female migrants usually carried out the kinwork and shouldered the burden of guilt and responsibility for such communication with, and caregiving to, parents and siblings (both their own and their in-laws).

'Neighbouring nations': New Zealand and Singapore

New Zealand migrants and their parents

> *In fact, we've probably maintained almost as much face to face contact now as we did when we were actually living in New Zealand. The distance is different, but the quantity of visits [is the same].*
>
> (migrant)

The geographic proximity of New Zealand and Australia, along with their common Anglo-Celtic roots since Western colonisation, no doubt contributes to the continuing and close social, political and economic relationships between the two countries. Long histories of easy movement of people between Sydney and Auckland became regulated in the unique 1973 Trans-Tasman Travel Arrangement (DIMA, 1997), which ensures that all New Zealand and Australian passport holders are entitled to travel freely between the two countries (Brosnan & Poot, 1986; Carmichael, 1993). In a pattern that is long established, most people who enter Australia from New Zealand reside on the east coast of the country; Sydney and Melbourne were most popular earlier in the twentieth century, with Brisbane and the Gold Coast becoming more significant settlement destinations from 1976 onwards (Pool, 1980). New Zealand Maoris also tend to settle in New South Wales and Queensland, although a substantial number have gone to Perth, Western Australia (Buetow, 1994, p. 312; Carmichael, 2001). While people of all ages are able to enter Australia from New Zealand, the majority is between the ages of 15 and 35, and aged migration occupies a particularly small proportion of trans-Tasman movement (Price, 1980). Indeed, New Zealand-born migrants aged 65 and over are markedly under-represented (Pratt, 1980; Carmichael, 2001).

People born in New Zealand have been travelling to Western Australia since at least 1891, albeit in relatively small numbers and often in response to mining-related economic booms in the remote regions of the state. This close alignment with mining-related employment contributes to the tendency of New Zealand-born to reside outside of the capital city, as compared with other people from overseas who tend to congregate in cities (Pool, 1980; Buetow, 1994; Carmichael, 2001). This industry-specific migration pattern is likely to contribute also to the consistently larger numbers of male rather than female New Zealand-born within Western Australia. Indeed, movement of people between New Zealand and Australia shows a very clear pattern of responses to economic situations in each country, with a highly mobile workforce tending to relocate as the economy improves in one or the other nation (Brosnan & Poot, 1986). Nevertheless, since the 1960s the movement has been largely in favour of Australia (Price, 1980), such that New Zealand became the most significant source country of immigrants in Australia in the late 1990s, outnumbering United Kingdom arrivals for the first time in 1996 (Jupp, 1998).

The New Zealand sample

A total of ten migrants from New Zealand were interviewed for this project (see Table 3.5). Unlike the other sample groups, all but three of the migrant interviewees were male. All were between the ages of 40 and 55, except for one who was in his 30s. Only two of the interviewees did not have either tertiary or skilled trade qualifications. At the time of interview, one person was unemployed and one person was working less than full time. Indeed, all except one reported their income level to be either 'high' or 'medium'. Only one of the interviewees was not of British descent, in spite of the fact that many New Zealand Maori and Pacific Islanders have settled in Perth in recent decades. Further attention to the Maori and Pacific Islander experiences of transnational caregiving is an important area for future research (but see Lee, 2003). The interviewees were all either married or in long-term *de facto* relationships; more than half had Australian-born partners, and the remainder had New Zealand-born partners, supporting Borrie's (1994) claim that the level of inter-marriage between the two countries is high. Only four had children, and in one case these

Table 3.5. New Zealand sample

	Migrants	**Parents**	**Other kin in New Zealand**
Male	6	2	—
Female	2	2	1
Couples	1	4	—

children had almost reached adulthood and independence. All of the interviewees had siblings living in New Zealand, but only four had siblings living in other countries around the world. Only one had siblings in Australia, who also resided in Perth.

The New Zealand interviewees appear to be highly mobile. Their ties to Perth are minimal, relying primarily on home ownership and employment, which interviewees said they would be happy to relinquish should a good enough reason arise for relocating. This sample characteristic repeats the statistical evidence that the New Zealand-born are highly likely to move within Australia after arrival (Guest, 1993; Carmichael, 2001). The lack of children, in particular, reduces the difficulty of moving to another city or country, as there are no education paths or childhood friendships to take into account in making the decision. Perhaps for similar reasons, this group of people frequently make return visits to New Zealand. Over half of the interviewees have been visiting every 12–18 months, one visits every two to three years, with only the remaining three making relatively infrequent visits back home. One of these has siblings in Perth, and has received visits from New Zealand family residents approximately every two years, possibly reducing the sense of obligation to visit. The other two have been limited in their ability to visit by their child-rearing obligations and the financial and time constraints associated with having children. All three stated their intention of increasing their visits to New Zealand in the future.

This entire group of interviewees is relatively young, and their parents in New Zealand tend to be in good health. Half of the interview sample had both parents living in New Zealand, which appears to significantly reduce the sense of a need to provide continuous support. In only two of these cases, fathers had some health problems and there was some concern on the part of the migrant children as to whether their mothers were coping with the additional need to provide care. However, locally resident siblings provided reassurance that this was not a significant source of concern. Eight of the 14 parents in New Zealand were in their 70s, four in their 80s and two in their 60s. In one instance, a father lived alone after the death of his wife, but the majority of the single parents were widowed mothers. Most of the parents had at least one daughter living in close proximity, able to help out when required.

All of the parents live in their own homes, two of them providing occasional accommodation support to adult children living in New Zealand. Only one parent couple and the single father receive regular care assistance, in the form of meals or cleaning services. All are living either within or in close proximity to a large town, with only one living in a capital city. In most cases, they reside in the homes in which they had raised their children, and thus reported having good knowledge of and easy access to social and support networks, activities and clubs, medical care, local government assistance and so forth. All except one had both the desire and the ability to make regular visits to Australia to visit their children in Perth. The most regular visitors are one couple who have made almost annual journeys to Perth since their child's migration in the early 1980s.

Migration to Australia

Almost all of the migrants in the New Zealand sample had arrived in Australia in the 1980s, with only one arriving in the 1970s and another two arriving in the 1990s. The reasons for moving to Perth fall into two categories: for approximately a third of the group, emotional reasons were the primary motivator – usually to follow love, but in one instance to pursue spiritual development. The remaining two-thirds of the group moved primarily for work-related reasons, either to have a working holiday, to pursue more lucrative or engaging employment opportunities, or to undertake further study for career development (see also Wood, 1980; Brosnan & Poot, 1986; Borrie, 1994). Only two of these work-seeking migrants indicated that a lack of opportunities in New Zealand contributed to their decision. For most, it was the search for diversity or adventure that encouraged them to seek employment outside of New Zealand; there was a sense in which Australia was perceived as a 'bigger pond' in which a career could be carved. Indeed, several indicated that their intention had been to move on from Australia to other international locations; however, this movement was often forestalled by an appreciation of the easy lifestyle in Perth. Almost all indicated that they would be willing to return to New Zealand to live permanently, if the employment opportunities were appropriate. Two were actively seeking opportunities to return, supporting the claim that return migration is an important component of migration flows between Australia and New Zealand, with people tending to relocate relatively easily between the two countries (Brosnan & Poot, 1986).

Due to the Trans-Tasman Travel Agreement, none of the migrants needed to seek visas or special entry permits in order to move to Australia. This is likely to be a key factor contributing to two interesting features of the New Zealand experience. First, there was very little resistance from parents and other family members in New Zealand to their decision to move to Australia. Most parents understood their children's travel to Australia as a common element of leaving home; indeed, more surprise was expressed at those children who had only travelled as far as Australia, because there was a common expectation that the 'OE' (overseas experience) would take young adventurers to Europe, America or at the very least to travel through Asia (see also Carmichael, 1996). Thus, licence to leave was not mentioned as problematic for any of the migrants, and none of the parents expressed any resentment or regret about their children living overseas. There was some reflection on the particularly long distance that Western Australia represented in relation to New Zealand, and some expressed a longing for their child to be living on the east coast, in Brisbane or Sydney particularly. However, even Perth was perceived as being within relatively easy reach, with airfares and telephone calls seen as accessible to most of the migrants and their parents. Second, again possibly because of the lack of formal steps required to enter Australia, none of the New Zealand sample thought of themselves as 'migrants'. In fact, when the term was used in interviews with parents and siblings in New Zealand, it was most often greeted with laughter and wonderment – the notion of 'migrating' to Australia was clearly odd and amusing from a New Zealand perspective. It also appeared to be odd from an Australian perspective. As one migrant said:

> *I was the invisible migrant here ... Most people never knew that I wasn't an Australian. Even the government didn't. Nobody. ... I never saw myself as a migrant.*[18]

Many interviewees put this down to the perception that New Zealand is Australia's 'younger brother' – there was a sense that migrating to Australia is as unnecessary as having your sibling 'adopted' by your own parents. It is perhaps for this reason that only half of the New Zealand migrants we interviewed have taken up Australian citizenship; yet even this constitutes a higher proportion than the national average. The 1996 Census indicates that only 32 per cent of New Zealand-born had taken up Australian citizenship, much lower than the average of 71 per cent for other overseas-born people in Australia (Carmichael, 2001).

There is no pattern in our sample for New Zealand dual citizenship in terms of year of arrival, absence or presence of children, gender, education, childbearing, income or employment. Most of those who did not have Australian citizenship attributed it simply to 'laziness'. For example, two mentioned that the relevant forms had been sitting on their desks for a long time, but had never been completed or posted. Two said that they felt they will never 'be' Australians, that they will always be 'Kiwis living in Australia', and so did not intend ever to take up Australian citizenship. No doubt the special consideration given to New Zealand migrants plays a significant role in this decision, in spite of the possibility of dual citizenship. One example supporting this speculation is that of a New Zealand migrant who had taken the steps to acquire Australian citizenship only after he decided to return to tertiary education. Citizenship was a necessary precondition to be able to apply for the deferral of university fees through the HECS (Higher Education Contribution Scheme) program, enabling applicants to pay for their university education only after they begin earning a substantial income. Access to HECS deferral is one of the few benefits available only to Australian citizens, and not to New Zealand residents in Australia. Indeed, prior to 2001, all New Zealand citizens had almost complete access to all other welfare benefits in Australia. However, new legislation means that now only 'eligible' New Zealand citizens can receive these benefits (see Burn & Reich, 2005, p. 491).[19]

The low take-up rates of Australian citizenship are possibly also a result of a tendency for New Zealanders to be 'treated as though they were internal rather than international migrants' (Jupp, 1998, p. 135; see also Price, 1980), with a relative absence of demands for New Zealand migrants to 'abandon their alien ways and become Australian' (Jupp, 1998, p. 135). Even apparently 'non-white' New Zealanders, such as the Maori population in Western Australia, report high levels of social and cultural acceptance in Australia (Jones, 1983).

Aged-care facilities and expectations

As is the case in all of our samples, in New Zealand older people are an increasing proportion of the population. The major statistics collection agency in New Zealand, Statistics New Zealand (2004), reports that the 65 and over age group rose from 9 per cent in 1951 to 12 per cent in 2001, with an expected dramatic rise to approximately 25 per cent by 2051. In 2001, 9 per cent of the over-65 age group were classified as 'old–old' (85 and over), a figure that is expected to increase to 22 per cent by 2051. Historically, women have been over represented among the over-65s. However, there are signs of a trend towards a greater gender balance. In 2001, only 53 per cent of those aged 65–74 were women. The older population tends to be concentrated in particular areas of New Zealand: predominantly Auckland, Canterbury and Wellington. The three regional areas with the largest numbers of elderly people are the West Coast, Gisborne, Tasman and Nelson. Our interviewees in New Zealand included people from most of these regions, with only two residing in other regions.

As was the case with the sample of parents interviewed in New Zealand, the vast majority of New Zealanders aged 65 and over in 2001 own their own homes (80 per cent of those aged 65–74) and live in private dwellings (98 per cent of those aged 65–74) (Statistics New Zealand, 2004). Indeed, the rate of institutionalisation for aged people in New Zealand is very low; in 2001 of the 2 per cent of people aged 65 and over who did not live in private dwellings, only 9 per cent lived in hospitals or nursing homes with the vast majority living in more independent forms of age-specific accommodation, such as retirement homes (Statistics New Zealand, 2001 Census). Most New Zealanders aged 65–74 live with a partner – 80 per cent of men and 59 per cent of women. However, women over the age of 75 are much less likely to have a partner (Statistics New Zealand, 2004). Of those people living in private dwellings, 43 per cent of those aged over 65 reported in a 1996 survey that they required some form of assistance from local services, and almost all indicated that they were able to access the facilities they required (Ashton, 2000). However, for elderly people living in rural areas, access to age-appropriate services and accommodation appears to be less readily available, requiring many to relocate to larger towns (Joseph & Chalmers, 1996).

New Zealand introduced an old age pension in 1898 and established a universal health care system in the Social Security Act of 1938, which guaranteed access to health services for all New Zealand citizens (Fougere, 2001; St John & Willmore, 2001). Policy debates throughout the post-war decades make it clear that, in terms of aged care, this system aimed to provide security for those who were not able to access services through other, preferred means (Uttley, 1995). For example, family members are expected to provide primary care to their elderly relatives; when they are not available, it is expected that local religious, voluntary and community groups would provide such care (Uttley, 1995). The full support of the state for elderly persons does not become active until this latter community option also fails. There is clear evidence that the government policy is also a cultural ideal for people in

New Zealand: it is generally understood that care for the elderly who need it should be provided out of a sense of devotion and social conscience, and should not be the object of profit (Ng & McCreanor, 1999).

The government provides a pension income for all New Zealand citizens over the age of 65 (an increase from 60 years of age in 1992) called New Zealand Superannuation, which is the main source of income for most retirees (Fergusson et al., 2001). This pension is not means-tested and is provided at a flat rate for all recipients regardless of their pre-retirement income (Fergusson et al., 2001) and regardless of whether they earn additional incomes through employment or investment after the age of 65 (Statistics New Zealand, 2004). The fact that this pension is not linked to previous incomes has been acknowledged as particularly important in reducing the inequalities that many aged women face in other countries (St John & Willmore, 2001). For those older people in New Zealand who have few assets and limited income, there is a range of additional income support services available, including rent assistance.

In the 1980s the social welfare approach to health and aged care was challenged by a shift towards the 'enterprise state' (Opie, 1992), evident in gradual de-institutionalisation (but see Uttley, 1995) and the reduction of government subsidies for health care and related costs. In the 1990s, the Health and Disabilities Services Act (1993) marked a major restructuring of health- and aged-care provisions; it brought an increasing emphasis upon a 'quasi-market' model of healthcare, which was believed to be necessary to procure appropriate efficiencies in health service provisions. However, the 1996 election resulted in a swing back towards identifying healthcare as a social necessity, and recent state policy has brought further reforms of the healthcare system that hark back to the prior emphasis on health as a public right, rather than a private good (Ashton, 2000). New Zealand, like the Netherlands, has a fully national program for providing long-term care to elderly people: services are provided free of charge to those who cannot pay themselves, but means-testing is applied so that those who are capable of paying for their care do so to the full extent of that capacity (Merlis, 2000). However, in New Zealand this state-provided care is still clearly understood to be a supplement to the assumed availability of 'community care', whereby family members are expected to provide most day-to-day caregiving needs with only minimal state assistance until or unless that is no longer possible (Opie, 1992). Since 1998, one central agency has been responsible for contracting service providers to provide all publicly funded health services, including institutional and home care services, so as to 'develop a comprehensive and cohesive set of services for older persons' (Ashton, 2000).

The lack of automatic state assistance for aged care is in spite of the fact that recent demographic shifts make informal care provision increasingly unlikely. For example, as is the case for other countries included in our research, couples are having fewer children who will be available to care for them, and they are living longer (Opie, 1992). While this might suggest some sense of emergency about future provision of aged care, it should also be acknowledged that the aged population in New Zealand, like that of Australia, is estimated to be only a small proportion of the total population when compared with other OECD countries – comprising an estimated 16 per cent in 2020, significantly less than the 20 per cent or higher proportions of population expected to be 65 and over in countries such as Italy, France, Germany and Japan by that time (Anderson & Hussey, 2000).

Singaporean migrants and their parents

> *As Asians, we are not like Aussies where at 17 you are told to move out of the house. We are Chinese; we are supposed to stay with our parents.*
>
> (sibling of migrant)

Western Australia currently hosts the largest population of Singaporean-born migrants (10,270), followed by the states of New South Wales, Victoria and Queensland.[20] This is a highly skilled group of migrants, with nearly 60 per cent of Singaporeans living in Australia holding some form of tertiary qualification. In Western Australia, 63 per cent of employed Singaporeans work in skilled occupations and more than 90 per cent speak English well or very well (ABS, 2001). Many of these Singaporeans initially came to Australia as international fee-paying students and had intended only a transitory sojourn for academic reasons.[21] As one parent said:

> *In Singapore when you don't have enough points to get into the university, you have to do something for your children. Either they come out to work, or if you have money, you send them overseas to study.*[22]

Many of these students later converted this temporary stay into a permanent residence because they found they preferred the relative social and political freedom of Australia to conditions in Singapore. One migrant commented:

> *When I came to Australia, everything fell together. I was never a good student. I didn't really study hard in university. When I went to*

Australia, I scored straight A's all the way! I was like, stunned. The whole family was stunned. How the hell did he do it? I think it just fell together. There was something missing that I found in Australia basically, I suppose.[23]

Migrants seeking permanent residency upon completion of their tertiary education in Australia make application through the Skills stream (see Chapter 2). Singaporeans who have studied in Australia are likely to be successful in their application; they have already spent time in Australia and have had an opportunity to become acculturated to the Australian way of life. In fact, as Quigley & Menon (2004, p. 256) report, between 2000 and 2001, over 50 per cent of migrants who were successful in their application using the MODL system (part of the Skills Stream) were international students.

The main circumstances motivating Singaporean migration to Australia are a combination of 'push-pull' factors that involve the practical (such as education and lifestyle) and the emotive (for example, a response to an intrinsic sense of not belonging to or, in some cases, rejecting Singaporean life and culture). These sentiments have been researched and documented in studies conducted both in Singapore and Australia.[24]

The Singaporean sample

The Singapore sample comprised 14 migrants resident in Perth, Western Australia and 12 relatives living in Singapore. Although the majority of this sample was of Chinese descent, other ethnic groups were also represented (including Malay, Sikh and Indian-Eurasian). Migrant participants were aged between 22 and 62 years, and had been resident in Australia from between 2 and 15 years (except for one woman who had been in Australia since 1968). Of the nine female migrants, five had completed tertiary studies in Perth; of these only one married a Singaporean. Of the four women who had not studied in Perth, three migrated with their husbands, and one came to marry an Australian. All but one of the five male migrants had also studied in Perth prior to their permanent migration; the other was a Singaporean businessman who ran businesses between Singapore and Perth. All have permanent residency in Australia while retaining their Singaporean identity, except for two who had taken out Australian citizenship (one of these has dual citizenship).

The Singapore-based kin interviewed for this study included three brothers (aged between 29 and 32 years) and nine elderly relatives (aged between 49 and 70 years, comprising seven mothers, one father and one

Table 3.6. Singaporean sample

	Age	**Gender**	**Social status**	**Living environment**
Migrants	22–62	9 F 5 M	Middle–Upper	House
Siblings (S) and Parents (P)	S: 29–32 P: 49–70	8 F 4 M	Lower–Upper	Flat, Condominium, House

aunt). Both migrants and Singapore-based interviewees came from a range of socio-economic backgrounds. This shows in their different living environments (including flats, condominiums and houses), living/household strategies and parental-care arrangements (with family/extended family or live-in maids). Of the elderly relatives interviewed, five lived with their spouses, three lived with adult children and one lived with a parent. Of the three brothers interviewed, one lived with his parents and two with their spouse. It is noteworthy that one father, although living in Singapore, has permanent residency status in Australia, while another parent couple is applying for permanency, to facilitate long visits to Western Australia. Table 3.6 sets out the main demographic descriptors of the participants.

Singaporean parents and their caregiving needs

Singapore society is ageing rapidly. Projections are that the proportion of the aged (defined as persons aged 60 and above) will increase from about 10 per cent today to 26 per cent in 2030, or one in every four persons. Additionally, the number of wholly elderly households increased from 10,400 in 1990 to 25,700 in 2000, at an average annual rate of 9.4 per cent. This makes Singapore the fastest ageing society in Asia after Japan (Lee & Yeo, 2003).

Due ostensibly to an acute shortage of land, elderly residential care in Singapore is provided through a host of government and charity-based nursing homes that offer largely hostel-style accommodation. Attending to the needs of both medically fit and less-able elderly, nursing homes in Singapore fall into a number of different categories: the main distinction is that between residential and non-residential facilities. Within this classification lie considerations of the medical needs of the elderly. Table 3.7 denotes the main classes of facilities offered within this system, which is perceived as having both positive and negative aspects. Interviewees saw positives in the variety of organised activities provided for the elderly, and the existence of elderly-friendly accommodation alternatives in the realm of public housing. As noted by one of the Singapore-based parents:

> *Actually, the Singapore government really takes good care of the senior citizens ... They have activities like singing, karaoke, and mixing around ... and they are building HDB studio apartments for senior citizens, either couples or singles. You either rent it out or you buy it, and they have all kinds of facilities ... Things like when they are in trouble, or when they faint or nobody is there, there's always a bell or something for them to press so that people will rush up to help them when they are in difficulties. They have everything in that studio apartment ... So I think in this way the government did a very good job for its senior citizens.*[25]

More critical views of the aged-care system in Singapore tend to focus on the incapacity of institutionalised care facilities to provide adequately for the social and emotional needs of the frail elderly. Both migrants and their parents see this as an issue of concern. Such frail-aged care institutions are compared to 'hospital' situations with inadequate facilities. They are described as 'dehumanising', 'grossly neglected', 'like going to a grave', and people in such 'old folks' homes

Table 3.7. Residential and non-residential elderly care in Singapore

	Residential	**Non-residential**
Medical	Nursing home: residential centres with medical facilities Respite care centre: catering for post-hospitalisation nursing care	Rehabilitation centre: catering for physiotherapy treatments
Non-medical	Sheltered home: residential accommodation for elderly, who require minimal medical attention	Drop-in facilities: catering for casual day activities (seminars, excursions) for retired elderly Day-care centre: non-residen tial day-care facility for elderly whose carers work during the day

Source: Home Nursing Foundation (http://www.hnf.org.sg/webtop/elderly.phtml, accessed September 2005).

seen as preferring to die rather than 'basically sit there, with nothing to do and no-one to talk to day in and day out'.[26]

Institutional care, then, is not an option any of the interviewees found acceptable for their own relatives. Migrants' and parents' sense of caregiving responsibilities usually centres on the Confucian notion of 'filial piety', which presumes that adult children (and especially adult sons) have an obligation to care for their parents (see Treas & Wang, 1993; Bengtson et al., 2000, Chapter 3). This concept of 'filial piety' was not restricted to the ethnic Chinese participants – migrants and their Singapore-based relatives of other religious and ethnic backgrounds shared much the same views. As one Singapore-based brother stated emphatically 'caring for one's parents or elders ... it is a must in Islam. Without a doubt. It is in fact an obligation'.[27]

As we discuss in more depth in Chapter 4, filial piety is expressed through obligatory financial support expected to flow from the younger to the older generation, while many parents also favour the idea of living with their migrant children post-retirement. This, however, does not apply in all cases. In some instances, the traditional notion of filial piety has found a new trajectory, with young migrants seeking out non-conventional methods of 'caring' for their homeland-based elderly, for example, by financing computer lessons to familiarise them with email facilities, or financing extended holidays to destinations of interest, instead of falling back on the time-honoured practice of putting their pay packets in the mail. Further, a number of parents have abandoned the idea of living with their adult children, instead preferring an independent lifestyle in a separate residence. Such independent living is feasible for those who can afford to engage the services of a maid to provide assistance with household chores and more intimate aspects of caregiving.

Currently, many households in Singapore, like in Italy, employ full time, live-in maids who play a significant role in child-care and the completion of household chores (cooking, washing, cleaning). Many of these maids come from other South-East Asian countries, including the Philippines, Indonesia and Thailand, and are made available to Singaporean families through a network of agencies based both in Singapore and the countries of origin. According to a number of parents, maids have also become an option in the care of the frail-aged. As one mother noted, 'the government allows us to employ a maid from foreign countries, so maybe this is the help I'm going to get'.[28] One of the migrants interviewed described the same approach to caregiving in the case of one of his uncles. His account indicates at the same time how financial support provided by adult children (including migrants) may be used – namely for the hiring of a live-in maid. In his words:

> *My uncle ... was bedridden, and then they had to get a Filipino maid. So the children had to provide and pay for the Filipino maid. That's the only difference. Here [in Australia] you have the Silver Chain that come to visit the old at home ... In Singapore, I don't think you've got that. You have to provide it yourself. So my cousin pays a Filipino maid especially to look after my uncle. That's the way I think they look after the old in Singapore.*[29]

Many parents, however, do not require such financial support; they have already planned for the post-retirement phase of their lives with savings plans and/or post-retirement incomes. This takes the form of government-provided pensions, and/or superannuation plans like the Central Provident Fund (CPF). The former is a system available only to government employees working in selected sectors (including teaching and nursing), while the latter is a mandatory superannuation system that includes regular contributions from both employers and employees, which is in turn divided into different components. These components are earmarked for particular purposes pre- and post-retirement, such as mortgage down payments, medical expenses (so called Medisave) or post-retirement income.

Refugees from Iraq and Afghanistan

> *My mother was very happy for me when I was granted a visa to come to Australia. She told me to, 'get on with your life' and not to worry about her.*
>
> (refugee daughter)

> *What kind of world is this, having only two daughters and none of them living with you?*
>
> (Afghan mother)

The complex stories of the Iraqi and Afghan refugees reflect their nations' histories of political upheavals. The individual responses to the resulting economic, social and political situations are extremely diverse, but share some common patterns. For example, their arrival in Australia usually followed a series of movements through several transit countries in search of some security and a reasonable standard of living. The particular migration trajectories of each family and individual have resulted in significant differences in the capacity of these refugees to provide care for their parents and other relatives overseas. They have also contributed to different settlement opportunities and experiences in Australia. Especially important in this context is to distinguish between those who arrived in Australia as 'authorised' refugees under the auspices of United Nations High Commission for Refugees (UNHCR), and those who came as 'unauthorised' refugees, mainly by boat and with the assistance of people smugglers (see Chapter 2).

Iraqi refugees and their migration histories[30]

Iraqi refugees of Shia Arab and Kurdish ethnic background have been leaving their home country for the last 40 years, mainly in exile to Iran. A considerable number of these had been expelled from Iraq at the time of the Iraq–Iran war; others fled at the time of the Gulf War in 1991. Of these Iraqi, a number subsequently migrated to Australia as authorised refugees. Their arrival contributed to a growth in the Iraq-born population in Australia from 2273 people recorded in the 1976 Census, to 14,109 in the 1996 Census, 20 years later. It is likely that the majority had fled to Australia in response to the Gulf War. Whereas most were Special Humanitarian entrants, some arrived under family and skilled migration categories. In 1996, of the total Iraqi-born population, 4.1 per cent were residents of Western Australia.

It was not until mid-1995 that unauthorised arrivals from Iraq began to enter Australia. Most of them had left Iraq after the first Gulf War for neighbouring countries such as Jordan, Syria and Iran. They wished to settle in a peaceful country permanently and sought the help of smugglers to find such a country. Shii people who had been involved in uprisings against Sadam Hussein also escaped; in their case to refugee camps in Saudi Arabia. Those who came to Australia had usually travelled from Iraq to nearby Iran or Syria, then to Indonesia, before they arrived as unauthorised refugees by boat on the northern coast of Australia. In 1999–2000, the total number of Iraqi-born Temporary Protection Visa (TPV) holders in Australia was 871; by May 2002 this had increased to 3082.[31]

Afghan refugees

Afghan migration history in Australia is in some respects even more complex than that of the Iraqi refugees. Afghan camel drivers were important to Australia in the second half of the nineteenth century, as they were central to the opening up of the remote northern and central desert to commercial activity (Jupp, 2001, p. 164). However, the White Australia Policy contributed to almost negligible numbers of Afghan-born in Australia from the early twentieth century, and this remained the case until relatively recently. From 1980, the number of Afghan refugees in Australia increased by an average of 500 per year (Jupp, 2001, p. 164). By the time of the 1996 Census, there were 5824 Afghanistan-born residents in Australia, mostly refugees.

Those who migrated prior to 1979 were in Australia as students or married to an Australian citizen, but the more recent migrations out of Afghanistan were in response to major political upheavals and civil wars that forced people to flee their homelands. In 1978–1979, when communists seized power under President Taraki and soldiers from the Soviet Union invaded Afghanistan, more than two million Afghan people became refugees, possibly the largest sudden refugee exodus in the world. Many of the refugees were once supporters of the guerrilla movement of Mujahedin, a collaboration of various ethnic groups, including the Tajik, Hazara and Uzbek, fighting against the communists and their Soviet allies. Those who migrated at that time were primarily middle- and higher-class professional people and intellectuals from the large cities of Kabul and Heart, who fled to nearby transit countries and then to a Western country.[32]

Throughout the 1980s, as the civil war between the Mujahedin and the communist regime intensified, more and more people began to flee Afghanistan. Some would return to their home country time and again in the hope of being able to live there in safety, but many ended up staying in Iran or moving on to other neighbouring countries such as Pakistan or India. They were often able to get valid visas during this time, and because they were well educated, those who sought refugee status in countries such as Germany, the United States or Canada did not have a great deal of difficulty being accepted by these countries. Australia also accepted small numbers of these Afghan refugees as part of the Special Humanitarian program.

In 1992, the anti-communist war led to the downfall of the communist Najibullah government and the departure of the Russian military forces from Afghanistan. Many Afghan refugees who lived in Iran and Pakistan decided to return to Afghanistan at that time. However, old hostilities soon resurfaced and triggered a civil war extending to all villages (*gharyeh*) and districts (*vela sovali*) of Afghanistan. The country plunged into lawlessness and became an arena of war between violent ethno-political oppositional groups, culminating in 1995 in the rise to power of the Islam-fundamentalist Taliban regime. This situation resulted in yet another major wave of migration, this time primarily of people who belonged to the Hazara Shii ethnic group. Unlike the earlier refugees, the Hazara were mainly village people, with only a few years of religious schooling (*maktab-e akhondi*), and usually a peasant background in farming and herding. They had lived in the most disadvantaged villages and districts for centuries. Nearly two million of these Afghan refugees arrived at the Iranian borders during the last three decades of political turmoil in Afghanistan.

The majority of Hazara Afghan refugees left Afghanistan after, and as a result of, the Taliban rise to power in 1995. At this time, the largest Afghan cities fell under the control of the Taliban or Pashtun and their local allies. These began to occupy towns and villages populated largely by Hazara and Tajik Shii ethnic groups. Hazara families reported being harassed on a daily basis, their houses were regularly searched for weapons, their young sons were taken to fight in the front line and women were forced to provide Taliban supporters with food and other resources. Initially, many young men went into hiding in nearby mountains to protect themselves and their family, hiding during the day and visiting family at night. Women and elderly relatives were often interrogated and asked to provide information about the young men in the families.

Many families then decided to send their men out of the area. To do this they had to seek help from local smugglers to get them to Iran or Pakistan, and from there sometimes via Indonesia on to other countries, including Australia. Some had no time to say farewell to parents, their wives and children, as they had to escape at a moment's notice. Their families often did not have any news of them for months after they arrived in Australia. The Hazara family members who remained behind after their men left were also often forced to flee their villages due to their fear of becoming the subjects of revenge by the Taliban regime.

Arrival of unauthorised refugees in Australia

The Afghan and Iraqi refugees who had fled their homeland with the help of smugglers in the late 1990s, and who subsequently arrived (usually by boat) in Australian waters, did not receive a warm welcome by the Australian government. After an often-arduous journey in boats that were hardly seaworthy, they were immediately arrested by Australian border protection patrols, placed in detention centres, and subsequently subjected to extensive interrogations to assess whether they were genuine refugees. Such detentions could be for up to five years, leading to considerable mental and physical distress (Kamalkhani, 2004, pp. 240–242).

As 'unauthorised' entrants, the best these arrivals could hope for was a TPV. Often men had made the perilous journey first, then faced years of loneliness without any certainty of ever seeing their families again, because of the TPV exclusion on seeking family reunion. This may be why some wives and children also attempted to journey to Australia, again with the help of smugglers. In 2001, the perils of such journeys were well illustrated by the sinking of a boat with Iraqi refugees between Indonesia and Australia. On that occasion, 146 Iraqi children, 142 women and 65 men were drowned (Kamalkhani, 2004, p. 240). In the same year the notorious Tampa incident occurred in which 350 men and 21 families with 43 children, mainly asylum seekers from Afghanistan, were rescued in the Indian Ocean from their sinking ship by a Norwegian vessel, but then not permitted to land in Australia. Instead, they were placed in detention on the Pacific Island of Nauru, and their applications processed from there. It took two years for the first applications of these Nauru detainees to be processed; 20 were granted a TPV in 2004 (Kamalkhani, 2004, p. 240). The Australian government introduced some changes in the TPV system in that year, enabling a total of 9500 asylum seekers (among them a number of Afghan and Iraqi) on TPVs to gain permanent residency in Australia.

The samples

For this project, interviews were conducted in Western Australia with members of 15 extended families, six of them Iraqi and nine Afghan (see Table 3.8). Interviews were also conducted with their extended family members in Iran, including three of Iraqi origin and twelve from Afghanistan.[32] The Iraqi refugees had been living in Australia for periods of between two and six years. The Afghan refugees from the early waves of forced migration are mainly professionals, who had come from large cities in Afghanistan and belong to the ethnic groups of Tajik and Sunni. They had generally arrived in Australia under the auspices of the UN Special Humanitarian Program and had lived in Australia from between nine to fifteen years.

Notwithstanding the length of time they had been in Australia, most had not been able to gain employment commensurate with their level of expertise and education. They included two families where the husband had been in Australia a long time and had sponsored a wife from transit countries (organised through arranged marriages using their female kin relations in Iran). However, Afghan refugees from the third wave, such as most of the Hazara, mostly entered unauthorised; they were often single men, or men with wives and children still in transit countries. They

Table 3.8. Refugee sample

	Iraq-born	**Afghanistan-born**
Interviews in Perth	6	9
Interviews in Iran	3	12

had been in Australia less than four years (in one case only four months), at the time of interview. They had not been able to gain paid employment, although some worked as casual labourers in the black market.

Some families from the early waves of forced migration had been able to gain acceptance in Australia as an extended family, including siblings with their spouses and married children. However, arrivals from the third wave were seldom allowed entry as an extended family group, unless they already had one family member with permanent residency in Australia. In this group of later arrivals it was much more common to find individual family members like a sister, daughter or wife without their extended family.

Parents and other relatives of the Australian-based Iraqi and Afghan refugees are themselves also refugees. We were only able to interview parents and other members of extended families who live as refugees in Iran, in the cities of Mashahd, Isfahan, Shiraz, Yazd, Qum and Teheran. About two-thirds of these people had been in Iran for many years. For example, one sister of a Perth-based Afghan woman had lived there for 15 years, since the Russian occupation of her homeland. Another Afghan woman, who had sisters living in Pakistan, Saudi Arabia and Australia, had been in Iran 21 years, and all her children had been born there. On the other hand, three of the Afghan families we interviewed in Iran had only been there since the time of the Taliban regime. It is important to emphasise, though, that many Australian-based Iraqi and Afghan refugees have parents, siblings and other relatives scattered throughout other countries. Some of their relatives had not fled to Iran, but to Pakistan and India; from those countries some had applied for refugee status in other western nations such as Germany, Norway, the Netherlands, the United Kingdom, Canada or the United States. Extended networks of these refugee families, then, stretch between various countries, rather than mainly between home and host countries, as is the case with the migrants in our study.

Caregiving needs of parents and other relatives in Iran

Except for those who have lived in Iran for a considerable number of years, the 1.5 million Afghan and approximately 500,000 Iraqi refugees who live in Iran today face a precarious future. In the case of people expelled from Iraq at the time of the Iraq–Iran war, only those who are able to prove family links to Iran have been granted Iranian citizenship. Others have found their citizenship disputed by both Iran and Iraq, rendering them virtually stateless.[33] For identification purposes these Iraqis and Afghans were initially issued with green cards, now with blue cards. Mainly due to the pressure of large numbers of refugees, the Iranian government has in recent years placed restrictions on employment opportunities, access to education and any other welfare provisions for refugees who hold such green or blue cards. Afghan refugees have no right to register property or assets in their name, open a bank account or enrol their children at government schools. In fact, in June 2004, the Iranian government instituted a payment of $25 (US) to those refugees who returned back home (covering transport expenses back to their villages).

This policy of repatriation is directed mainly at the Afghans as the numerically largest group of refugees. It is most detrimental to Hazara refugees, who have a visible ethnic phenotype and are therefore easily identifiable, risking being arrested in the street, or at work, for any minor infringements. Behind much of Iranian government policymaking since the fall of the Taliban regime in Afghanistan is the demand that Afghan refugees return to their homelands. In mid-2003, all Afghan residents were asked to re-register with the authorities and to hand in their refugee blue cards, receiving only a three-month temporary residence permit instead (Strand, Suhrke & Harpviken, 2004, p. 3).[34] Similar pressure to return to their homelands is exercised in the case of Iraqis, although such a return is often unrealistic due to the continuing political instability in Iraq.

Given this set of issues, it is not surprising that the relatives of Australia-based Iraqi and Afghan refugees who we interviewed in Iran live in very difficult circumstances. There are no government provisions for the health and age care of elderly refugees, and limited employment opportunities for younger members of their extended families whose households they share. The work that is available to men is usually illegal labouring work on building sites or farms, while women with no male supporter or from low-income families usually take jobs as servants in Iranian families. Housing conditions are poor; and it is common for a family of four or five members to live in a one- or two-room house or flat.[35] At the same time, Afghan refugees in particular live under constant fear of being forcefully repatriated to their homelands. The fact that they experience such hardship is a great burden for their Australian-based family members, who know that without their ongoing financial support their Iranian-based relatives will simply not survive. Overseas relatives also carry the costs of unexpected hospital or other medical charges for their elderly relatives in Iran. Further, knowing that their relatives may be repatriated at any time, leads to persistent efforts on the part of Australian-based refugees to sponsor applications for migration to Australia. At the same time, Iranian-based parents are grateful that their children have been able to escape, even though this means that they are left behind, alone and often destitute. We return to these issues in Chapter 4.

Conclusion

In this chapter we have introduced and organised the discussion of our research data according to national groups. However, nation of birth is not the only, nor the main, distinction between the participants in this study. Indeed, we have highlighted some of the intra-sample variations, including gender, age and class, in these summaries. It is also imperative, at this point, to emphasise some key differences that cut across each of the national group samples. Clearly, not all transnational migrants, or their families, are the same (see Harney & Baldassar, 2007), and they do not differ only according to national or ethnic backgrounds. Following Grillo's (2007) critique of the tendency to homogenise the category of transnationals (see, for example, Appadurai, 1991), our findings indicate various kinds of transnationals and the types of transnational migration they are engaged in.

The majority of our sample – approximately 75 per cent – is comprised of professional, well-educated, relatively affluent, highly mobile individuals who could be described as 'individually-oriented', in that their migrations were primarily undertaken to pursue personal opportunities for employment, adventure, lifestyle or love. In addition, they are not connected to chain migration or ethnic community networks. The remaining 25 per cent (including all of the humanitarian migrants and about 10 per cent of each of the other samples) are perhaps better defined as 'working' or 'proletarian' migrants. They could be described as more 'communally-oriented', in that their migrations frequently formed part of family economic strategies to improve the wealth of both migrants and their kin who remain behind. Further, these migrants are often strongly associated with ethnic communities in the host country.

The professional, 'individually-oriented' migrants arguably have many of the characteristics collectively defined in the literature as 'cosmopolitanism'. The best way to define 'cosmopolitans' is debated in the literature (see, for example, the critique by Skrbis, Kendall & Woodward, 2006). One useful definition in the context of our research is provided by Hall (2002, p. 26; but see also Cheah & Robbins, 1998; Vertovec & Cohen, 2002):

> Cosmopolitanism requires the ability to draw upon and enact vocabularies and discourses from a variety of cultural repertoires. The cosmopolitan has the technical and intellectual resources or 'capital' to gain employment across national boundaries, and typically has an ability to traverse, consume, appreciate and empathise with cultural symbols and practices that originate outside their home country.

One problem with this distinction is that it implies that all professional migrants possess 'cosmopolitan' skills and sensibilities – very much like Kanter's (1995) world class citizens – while all 'working-class' migrants do not (see Werbner, 1999). Werbner's (1997, p. 12) distinction between translocals and cosmopolitans raises a similar issue. Werbner (1997, p. 12) defines translocals as 'those whose loyalties are anchored in translocal social networks rather than the global ecumene'. Grillo (2007) describes this division as fundamentally to do with class; the translocals are the proletarian version of cosmopolitans, 'concerned with ordinary, everyday activities of people operating simultaneously in, across, between, more than one nation-state'. Yet we found that all migrants, refugees and their families are engaged in the quotidian concerns of caregiving, regardless of their class background.

Such distinctions of class are nevertheless significant and we return to this discussion in Chapter 8. Indeed, the class divisions that cut across our samples are as important to consider as other cultural differences that tend to be associated with particular national groups or ethnic backgrounds. We attempt to keep all of these inter- and intra-sample differences in mind as we present our analysis of findings on transnational caregiving within the remainder of this book.

We also organised our discussion of the groups in this chapter according to the geographic proximity between migrants and their parents abroad. This is not to suggest that distance is the most important factor in transnational caregiving. In fact, we will show in the following chapters that the *capacity to care* (helped or hindered by macro- and micro-factors faced by each sample group, including the impediments created by national borders) together with the *sense of obligation* (due to cultural values) and *negotiated family commitments* – all concepts we introduced in Chapter 1 – are more likely to explain practices of transnational caregiving than geographic distance alone.

4
Transnational Caregiving between the Generations

> *When I first came to Perth my Mum was devastated. I said 'look mum, I'm still going to support you'. She was still thinking monetary-wise. She didn't think that my brothers could give the care that I'd given her since Dad died. She thinks that she is not going to get that sort of care if I get married, so it was a big issue culturally. I said to [my husband], 'if you marry me, I'm not asking you to take the responsibility, but you are married to a family you know, it's not just me. [My mum] has been depending on me, you know'. She was very devastated, she knew that eventually I have to be with my husband, that's what you get married for, and we wanted to start a family. It was hard.*
>
> (Singaporean migrant daughter)

This chapter develops the core themes of our inquiry. It addresses and expands the question, posed before us by Finch and Mason (1993, p. 162), of 'how significant are kin as sources of practical and financial support', and applies it to a transnational context. Do people feel they have a responsibility to provide such assistance for relatives, or at least to certain relatives across distance? Central to Finch and Mason's theoretical approach is the argument that family responsibilities are the outcome of *negotiation*. Caregiving between family members is not a straightforward product of fixed rules of obligation, but the result of longstanding processes of negotiation based on a combination of normative guidelines and negotiated commitments. The process of negotiation:

> Can only be understood with reference to the biographies of the individuals involved and the history of their relationships, as they

> have developed over time. Biographies are themselves part of the negotiating process.
>
> (Finch & Mason, 1993, p. 79)

The sense of normative obligation is often strongest between parents and children. When sense of obligation is less strong, as it might be for example, between siblings, then personal liking takes on greater importance. Thus, who will provide what kind of support is a matter of negotiation in practice, discussed sometimes in 'family conference', in other cases based on tacit understandings that flow from past practices, previous commitments and joint histories. What is at stake in such negotiations as to whether to provide goods and services to kin, are not only the material values of the support, but also the identities that are formed in the process, such as a 'reliable son', a 'generous mother', a 'caring sister'. Some family members are in a position to advance 'legitimate excuses' (Finch, 1989, p. 210; Finch & Mason, 1993, Chapter 4) for not offering support.

In the process of negotiation regarding the fulfilment of family obligations, the most important principle is that of *reciprocity*. If there is an imbalance, people feel subordinated and 'beholden' to one another. Even when an older person requires extensive personal care and is in that regard dependent, family members often use various strategies to ensure that the aged person does not feel beholden to others. However, this does not necessarily mean that any service given must be returned immediately and directly. The reciprocal gift may come much later or be given by another person. Finch (1989, pp. 165–166; Finch & Mason, 1993, pp. 51–57) uses the term '*generalised reciprocity*' to describe situations when people feel an obligation to pay back to a third party, even to the community.

Finch and Mason's research took place in the United Kingdom, primarily among families of what are commonly held to be 'Anglo-Saxon' origins. This means that they did not explore how socio-cultural groups differ in their practices of mutual family support. Also, they have not given consideration to transnational family relations and obligations. We are able to do both, namely draw comparisons between the caregiving practices of a number of socio-cultural groups *across* distance and national borders. Applying Finch and Mason's concepts cross-culturally and transnationally raises several specific questions. What type of caregiving and support do elderly people in different cultural groups require and receive from their families, and to what extent do their transnational migrant and refugee offspring contribute to such care? Do elderly people reciprocate when they receive care and support? If so, are transnational migrants and refugees recipients of their parents' support or is parents' support limited to 'local' children? And what types of support are most significant in each case? We develop these issues in depth below. We begin with an exploration of the two types of transnational caregiving that can be accomplished from a distance: economic exchange and emotional and moral support. We follow these with an exploration of accommodation, practical support and personal care, which in most instances can only occur during visits.

Types of transnational intergenerational support

Economic support

Material support between the generations, 'giving and lending money for specified purposes' (Finch, 1989, p. 15), may involve money transfers, gifts, assistance in finding work and the promise of inheritance. As we have noted in Chapter 3, the parents of most of the *migrants* included in our study, regardless of their national origins, enjoy a reasonable standard of living and adequate housing. They are thus not in need of ongoing material support and our findings show that transnational migrants (except for the Singaporeans and post-war Italians) do not see themselves as having major responsibilities in contributing to their parent's economic wellbeing, although they may desire to contribute support on special occasions. The situation is drastically different in the case of *refugees*. Their parents, and other members of their extended family who escaped from their homelands and now live in transit countries, are in dire economic circumstances, and the majority of refugees have a major commitment to helping these family members economically. We will begin our account with a discussion of their plight.

The provision of material support to family members is the most important form of caregiving among *refugees*. Most of the Perth-based refugees from Iraq and Afghanistan we interviewed (in fact two-thirds of them) spend a lot of their limited income on financial support to their parents and other relatives who are living in subsistence conditions in transit countries such as Iran. They do this sometimes for many years and at great personal cost. An Afghan man said on this issue, 'when I came to Australia I had to start work to earn money immediately. I had all my brothers and my wife's relatives to support back in Pakistan. I had to work for cash money and take whatever job I could'. When asked how much help he is able to give, this man said:

> *I send from my own income, with support of my brother, $500 to our relatives, and my wife from her own money $500 to her relatives in Pakistan*

> *every month. We are not able to help all of our relatives. This is the normal money we send but if they get sick, or have an operation or [need] medication, we are asked to send more.*[1]

They had done this for the last ten years.

An Iraqi man, on his own in Australia after having paid smugglers 10,000 Australian dollars to get to this country while his wife and children are still in Iran trying to get visas, told our interviewer that he sends them 400 American dollars every two months. He is on social security, but somehow has been lucky to find work in the black market with a farmer who employs refugees on a casual basis for six [Australian] dollars an hour.

Relatives who receive money include parents, brothers, sisters, grandparents and occasionally aunts, uncles and cousins.[2] The money is used for the bare necessities of everyday life, but also in some instances to sponsor a parent or sibling to come to Australia. One Afghan man sends money on a regular basis to his mother in Iran. But his ardent wish is to bring her out to Australia, because 'in our culture a mother would rather live with her son'.[3] An immigration agent has told him that sponsorship of his mother would require a deposit of 15,000 Australian dollars.[4] In order to deal with these demands, this refugee, who only works part-time, has decided to sell his house in Perth. This is a very drastic step, and in fact he is one of very few who had such an asset to sell in the first place. But refugees may try to sell other things. An Iraqi woman, who is desperate to visit her family in Iran, has tried for some time to sell her gold jewellery to other women in her community. So far she has been unsuccessful, she thinks due to the fact that 'Iraqi women don't want to spend their money; they need it for their relatives, either for a ticket to go and visit them, or to send them money to be able to support themselves while living in limbo'.[5]

Refugees often receive urgent phone calls from relatives pleading for money. Such pleas can lead to mutual embarrassment, because their sense of cultural obligations to kin presumes that such requests are met, even if the refugee cannot afford it or the request is inappropriate. An Afghan man commented,

> *I have relatives in Iran, Afghanistan and Pakistan. They all expect my support. Once my uncle's son asked for money to buy a video for his son. I have no video myself. They think that we who live in Australia earn a lot of money. They don't know how difficult our life is.*[6]

But the pleas continue, and most of the time refugees will try to respond. This is clearly in part due to 'generalised reciprocity'. Many of the refugees we interviewed had themselves been in desperate need for money at some time, to pay smugglers to escape from their homelands, to finance the sponsorships and airfares needed for their relatives to join them in Australia, or even in some cases to travel back to Iran to find a wife.[7] Often relatives or friends often stepped into the breach with a loan or a gift of money. Parents and other kin who receive material support are usually not able to reciprocate (although they may try in minor ways by sending small gifts in return for the economic support they receive). But we may assume that parents and other members of their extended family had given these refugees material support when they were children, and it is characteristic for Iraqi and Afghan adult children to in turn offer protection and support for their parents and other kin when they are in need. This sense of obligation is exacerbated particularly as there are few, if any, alternative sources of support from either the relevant states or local communities for parents to fall back on.

Compared with the economic support offered by *refugees* to their parents and other relatives, any financial aid given by *migrants* to their transnational parents pales in significance. While many migrants of the post-war era and earlier sent remittances to their family back home, generally only in the case of Singaporeans do the migrant children in our study who arrived in recent years contribute financially to their parents as they age. It is to the *Singaporeans* that we now turn. The stories they told in interview give a clear sense of the degree and type of reciprocity they expect in their caregiving relations. The exchange began in most instances with substantial financial support from parents to their children when these moved to Australia for their tertiary studies – a common aspect of life for Singaporean families. Typically children would live in properties belonging to their families, and would receive sums of money for their everyday expenses. It was usually the father who decided on this support. It is worth remembering the extensive business connections between Singapore and Perth. This makes it especially attractive for fathers to lend support through the purchase of a property that would then be available to their offspring during their studies and after.

At the time of interview all respondents from Singapore had settled permanently in Western Australia and were financially secure enough not to be in need of parental financial support. In fact, they had generally reached a stage in their lives when, as some said, 'we will try to give something back'. Several had started this process financially by paying for expenses when their parent(s) went on holiday, and buying airline tickets and also medical insurance for parental visits. This would occur most frequently when the parent was a widowed mother. As we will see later, such payments for incidental expenses (by migrants for parents and vice versa) also occur in other migrant groups. But more importantly, the majority of Singaporean migrants whose parents are still in paid work say that they expect to provide them with some financial support once they retire. This is their plan, even when parents are said to be comfortably off. For example, a migrant daughter, whose parents are very well off, felt that once they retired:

> *this is an issue I have to discuss with my brothers and sisters ... probably we will all contribute a little to their daily expenses and to their medical fees ... and make arrangements with regards to traveling, like airfares.*[8]

A mother also said that she and her husband, 'don't need any help from them [their two children] ... we are comfortable ... but my son is very responsible. He insisted on giving us something. So ... every month from his account he sets up an automatic deduction and posts it to my account'.[9]

It is clear from such comments that sending back money on a regular basis in the case of these Singaporean migrants is usually not a financial necessity, but rather a symbolic representation of the adult child's cultural obligation to care for parents. It is not the actual money that counts but the symbolic gesture implied. In that sense the gift of money to parents would seem to be more important in the Singaporean case than other caregiving practices. For Singaporeans, unlike for the other migrants in our sample, the provision of emotional and moral support appears to need to include financial support. A Singapore-based brother of one migrant demonstrates this when he said in a disapproving tone of voice:

> *The disappointment was when my eldest brother went to Australia. Some may see it as a selfish reason. But he is doing it for his family [wife and children]. Maybe his way of taking care of his dad and the rest of his siblings is calling us, talking to us on the phone, coming back to visit. I mean he does not give our parents money.*[10]

Generally, parents of the other transnational migrants who live in Italy, Ireland, the Netherlands and New Zealand were – like the Singaporeans – economically well off. This does not mean that these parents are all equally well provided for; some indicated that they have to live frugally. But most are reasonably comfortable due to their own savings or to state pensions. They include many mothers who are widowed and have adequate pensions or incomes from their late husbands. There is, then, no need for transnational migrants to contribute to these parents' economic support. *Italians*, who emigrated to Australia in pre-and post-World War II periods from an impoverished homeland, assisted their parents and extended family back home with regular remittances. Today this is no longer a common occurrence, although some parents receive financial assistance from their local children, particularly if they are living with them or close to them, as is often the case in Italy and Ireland.

Only one Irish-born migrant sent regular remittances (of about 25 dollars a week) to her mother. She did this during the last three years of her mother's life.[11] When asked whether they provide financial aid to their parents, most of the other migrants made comments such as 'my mother is well provided for' (Italian-born daughter)[12] or 'no need to help mother financially, because she has a very good income' (Dutch-born daughter).[13] And parents would make comments such as 'I would not expect financial support from them' (widowed mother in Ireland).[14] In some instances, especially after the death of their fathers, migrants' responses suggest that the issue of financial support has been considered, but so far there has been no need. An Irish-born migrant said about her mother, 'I constantly ask, "Are you OK for money?" And she says, h "I am fine",'[15] while a New Zealand-born man commented about his widowed mother, 'I keep thinking I am going to have to do something to financially support her, but the reality is that when I check out her financial situation, talk to her accountant ... I can tell she is OK with the money'.[16]

When the Italian, Irish, Dutch or New Zealand-born migrants we interviewed give any financial support to their parents, it is primarily to enable their visits to Australia. In other words, they pay for airfares. The opposite, however, is more common: parents paying for the airfares of their migrant children so that they can visit back home. Beyond such expenditures, made to satisfy both parents' and migrants' need to see each other, major gifts or loans of money appear to take place exclusively from parents to their migrant children. A number of parents across all cultural groups helped their migrant children with the purchase of a house; one Italian-born migrant received a large sum to pay for her wedding, and another to buy a car; a Dutch mother sent money to her daughter when she was unemployed, and another supported her daughter financially during studies. Parents would also send gifts of money for birthdays and Christmas, and give cash to migrants during visits. One Irish mother gave all her children a thousand pounds 'out of her will' to celebrate the millennium. Parents would also make comments such as 'if he ever needed help, a mother always helps out'.[17] And many migrants (from all sample groups) indicated that if they needed help from their parents they would only have to ask. Such financial support was generally given fairly and equally to transnational and local children alike.

There are two major exceptions to this in the Irish and the Italian case; these have to do with *inheritance*. Italian parents may give large gifts to their migrant children, but it appears that these can sometimes take the place of the inheritance that is left to local children. We found that many Irish parents had given pieces of their land to local children so that they would be able to build their houses in very close proximity to their parents. Italian parents also often left their homes to a local child, particularly the one they expected to live with during their old age. Transnational migrants of course miss out on such a valuable gift, as shown in the following comment by an Irish-born migrant, one of several siblings who had migrated to Australia:

> *My father gave all of them their land for their house, and they did not have to pay for it. We did not get anything, ... if they asked me for help I would have to think about it, I would always feel I did not get any help when we needed or wanted it.*[18]

There is one way of not missing out for such transnational migrants: *repatriation*. One widowed Irish father said that he always has a site on his property if his migrant daughter ever wants to come back. An Irish widowed mother explained that she had made her will prior to her most recent trip to Australia, but was not quite satisfied yet with her arrangements, because she would like to know whether her migrant daughter wanted land for building. In saying this she clearly had in mind that her daughter would return to Ireland. Several Italian parents commented that their migrant children could have their house and other valuable items 'if they returned' – this appears partly a kind of bartering technique to try to put pressure on the children to repatriate as the following quote by a migrant woman makes clear: 'They were very upset and they would always try to make me come back, subtle things, like, "If you come back I will give you this set of plates!", others, not so subtle, like, "When you come back, this [antique furniture] is here for you".'[19] There were also a few instances in the case of the Irish and the Italians, where parents assisted in finding work for a migrant child (more specifically for a son-in-law), as a form of economic support when migrants came 'home' for long-term stays.

In this context, it is worth noting that only the Irish and the Italian interviewees raised the issue of inheritance as a form of economic support. As mentioned, some Irish migrants questioned gifts of land to local children versus cash to migrant offspring; in other words, the issue for them was whether transnational migrants get their fair share of the inheritance. But one migrant from Ireland and several from Italy indicated that they are willing to *forego* their inheritance if their siblings at home provide or pay for their parents' personal care. They are concerned to rectify what they consider an imbalance in the extent to which they can provide personal care by relinquishing receipt of material rewards.

In overview, then, transnational economic support to parents is most important for refugees and Singaporean migrants, in the former case because of the difficult circumstances of their parents' lives, in the latter as a symbolic representation of filial piety. For other transnational migrants and their parents, the provision of economic support is more a matter of choice or practicality, being given intermittently in the form of loans or gifts of money for purchase of a house, maybe a car, help with educational expenses and particularly the purchase of airline tickets, either for parents who want to visit or migrant children to return home. In these cases the most substantial economic support flows from parents to their migrant children rather than the other way round. Thus, Italian, Irish, New Zealand and Dutch parents provide lifelong economic support to their transnational migrant offspring. While in many cases this appears to occur without reciprocity of a financial kind, other forms of reciprocity do occur. We return to this issue at the end of the chapter. This being said, it must be remembered that in a minority of cases, no economic exchanges occur. In these cases, parents would make comments such as 'we brought our children up to stand on their own two feet, to run their own lives',[20] and migrants said things like 'I did everything on my own, I have always paid my own way.'[21] Such minority cases occur in all cultural groups, even the Singaporeans.

Emotional and moral support

Finch (1989, p. 33) defines emotional and moral support as activities that involve 'listening, talking, giving advice, and helping to put their own lives in perspective'. We believe it is warranted to say that such emotional and moral support, including 'just being there for each other' and expressing one's love and sense of loss, are the *bedrock* of transnational family relations. Emotional support helps migrants cope with homesickness, and parents with their profound sense of loss about the long distance separating them from their children and grandchildren. It involves mutual comfort when crises occur due to illness, death or other family break-up. From a distance, emotional and moral support is accomplished through letters, phone calls, emails and other IT-based communications. We will discuss in Chapter 5 how and when each of these forms of communication is employed. Chapter 6, in turn, will cover the specific relevance of visits to emotional and moral support.

In general, our findings on emotional and moral support concur with those of researchers such as Bengtson and colleagues (Bengtson et al., 1994; Bengtson, Rosenthal & Burton, 1995) in their work on intranational family relations. We found that transnational migrants, once they are married and have children of their own, strengthen their bonds with their parents, and that many parents as they age (and especially after divorce or the death of their partners) come to depend increasingly on their children's support, including that of their migrant offspring. As found also by Rossi and Rossi (1990), some migrant sons have especially strong bonds with their widowed mothers. But this is by no means always the case, because the likelihood of such bonds is influenced by other factors, such as the number and gender of local children, and most importantly the cultural expectations regarding the caregiving responsibilities of sons versus daughters, together with birth order in the case of sons.

For a detailed account, we focus on cultural differences as expressed by our interviewees in the extent to which members of the various sample groups express emotional and moral support during the stages of their family lifecycle. Finch and Mason (1993) make an important distinction between *routine support* (mainly between mothers and daughters; and between siblings), and *crisis support* that is more a matter of negotiation. We have found such distinctions useful for our analysis, but we have added a third set of contexts that demands extensive emotional support, namely the *time of emigration*, and the immediate period thereafter. Although this can be seen as a type of crisis support, we deal with it as a separate issue because the stage is unique to transnational families.

It is instructive to begin our discussion of emotional and moral support with the experiences of the Irish and Italians in our study. Three-quarters of the Irish, and nearly all of the Italian migrants indicated that their parents had expressed strong opposition to, or extreme distress over, their decision to migrate. They were 'devastated', 'not supportive', 'very upset', told their child 'you are doing the wrong thing' and in several cases objected to their child marrying an Australian. This was sometimes accompanied by a stated belief that the migration would only be a temporary move. As one Irish migrant said about her parents, 'they still think it is a temporary move although I have been here 10 years'.[22] Similarly, many Italian migrants, like the daughter quoted below, recounted that their parents continue to express disapproval at their decision to settle abroad:

> *absolutely every single time we speak they have to comment, ma perchè sei andata a vivere così lontano [why did you have to go and live so far away?] ... They still say that. Every time we speak, every time ... It's a big issue ... They are still shocked.*[23]

Notwithstanding (or maybe because of) these strong negative emotions on the part of their parents, many of these migrants said they had experienced excessive homesickness and a sense of powerful loss being away from their extended family and especially from their mothers.[24] Many parents visited quite often, but to resolve their sense of homesickness, migrants – especially the Irish – would inevitably return home to the fold of their family, often for very lengthy periods. In fact, two-thirds of the Irish migrants returned for periods of two to four years (in one case as much as 15 years) before settling again in Australia. Both Irish and Italian parents strongly encourage the return of their migrant children, for example, by finding work for the husbands of returning daughters as a means of keeping them close.

The reaction of Irish and Italian parents to their children's migration appears to indicate that they see it as unthinkable that anyone would choose to live away from their parents. It is important to remember in this context that they are usually surrounded by local children and so are not bereft of companionship or support. One Irish migrant noted this contradiction when he said to his wife, 'your mother still has seven with her, but she still was not prepared to let you go'.[25] Some Italian parents nevertheless complained that they would not have anyone to look after them in their old age, using this emotional leverage when they were actually still in very good health; as one migrant described, 'they had a couple of family councils and there were several attempts to guilt-trip me into coming back – "what will happen to your parents when they grow old and they need their daughter to look after them?"'.[26] Such expressions of anguish suggest that these parents have not given their migrant children 'license to leave'.

At the time of, and immediately after migration, extensive long-distance communication and visits by parents are generally not enough in the case of the Irish and Italians. It is essential for their sense of 'emotional healing', that they are able to make return visits. In the case of the Irish, this applies especially to young mothers returning home with their newborn baby to be 'pampered'. When not together, frequent phone calls and letters keep things going, although it appears that for the Irish and Italians, *routine* emotional support between parents and their transnational offspring also requires periods of close face-to-face contact during visits. It is possibly not surprising therefore that family *crises* frequently involve return visits. We return to this when we discuss the issue of personal care.

We found quite a distinct contrast between these two groups and the Dutch. While Irish and Italian parents let their children know in no uncertain terms that they were distressed about their plans to migrate, on the other hand, the Dutch parents in our study seldom conveyed to their children that they were upset about the decision to leave. This was notwithstanding the fact that these Dutch parents do not have the benefit of a large number of local children for their day-to-day comfort, because they generally have smaller families. During interviews, Dutch parents expressed a strong sense of sadness and loss, and were often tearful, especially about the absence of their daughters. In a number of cases daughters had migrated (generally accompanying husbands, or to marry an Australian), while sons had stayed in the Netherlands. As in the Italian and Irish samples, some of these sons were very supportive, but they could not quite satisfy their parents' emotional need for their daughters. Unlike the majority of the Irish and Italians, Dutch parents had clearly not been able or perhaps willing to communicate such feelings of sadness to their migrant children. One woman said, when asked what her mother's reaction was when told about her plans for migration, 'absolutely nothing, she said nothing. She has said nothing else about it, and we have never spoken about it again'.[27] The daughter was clearly disappointed by this. That Dutch migrants hoped for a more emotionally expressive response from their parents is also illustrated by the following comment:

> *When we decided that I would migrate I told my parents. And they had something like ... "Oh, very good, then we will have a holiday destination" ... and I was a little bit disappointed, like, is it that easy for them to see their daughter go off to the other side of the world? But they are very matter of fact, so it is very hard to find out what they really think ... and*

> *they won't show emotions very easily, also to make it easy for me. While I had something like, well, I would like to see some emotion, to see that they do care about me, something like that.*[28]

Dutch parents, then, do appear to give their migrant children license to leave, but many are actually unable to let their children know how much they miss them. Migrants, in turn, are often homesick (especially those who migrate by themselves to marry an Australian), but find it difficult to communicate the full extent of their misery and loss to their parents. The woman we quoted above said about her first year in Australia:

> *I was very lonely. And that meant that I phoned less, because I did not really want them to know, because it looked like I was failing, and I did not want them to know that I was feeling so bad. I could not talk about emotions that easily myself, just like them, and well ... what good would it do if they were worrying about me, while they were over there. So I would not call all that much, and they would not call all that much either.*[29]

The mother of another migrant said of her daughter, 'she had a very difficult time in the beginning. We did not know everything, because she did not write about it. I am glad I did not know, because I had enough sadness to cope with the first two years'.[30]

Common among all groups, but expressed in particularly strong terms by the Dutch, was a desire not to burden each other with negative emotions. In other words, the tendency not to communicate emotions such as sadness or worry was in itself seen as a form of emotional and moral support. Regular correspondence and phone calls, and especially the chance to visit migrant children, eventually helped parents come to terms with their sense of loss. As one Dutch mother said: 'you get used to it'.[31] But the inclination to be somewhat secretive did not stop. A Dutch migrant kept hidden from her parents for some years that she had separated from her husband, in order not to upset them. Several parents did not tell their migrant children about major health problems or a death in the family. One Dutch mother said about this, 'when I had a breakdown, you don't want to put a burden on them ... I don't want to worry them ... and I absolutely don't want to be a millstone around their necks'.[32] Such tendencies to secretiveness and restraint in emotional response occurred notwithstanding the fact that Dutch transnational families maintain extensive and frequent communications with each other through letters, phone calls and visits. They talk at length about everyday matters, but problematic issues are often not communicated. The tendency to be secretive (especially concerning health problems) occurs in all of the other sample groups as well, but the inability to express feelings of loss and anguish were particularly pronounced among the Dutch. We return to the issue of secrecy below.

As mentioned earlier, writers such as Rossi and Rossi (1990) suggest that parents become more emotionally dependent on their children as they age. Our research shows that Irish and Italian parents (and especially mothers) express a strong emotional need for their children, including their migrant offspring, throughout their lives. In a sense they never let go; this means there is hardly an increase in expressed emotional 'dependence'. In the Dutch case, on the other hand, such progression is quite apparent. Dutch parents find it difficult to express their emotional needs in the prime of their lives, but they seem much less willing and able to hide their sense of loss and emotional needs as they age. One Dutch migrant observed, 'I can't have a conversation with my mother on a weekly basis or she will say, "now at the end of my life, I need you so much, and now you are not here".'[33] Such comments suggest that the stoicism and restraint shown by these parents earlier in their lives are not maintained in their old age.

We dwelt at length on these three groups to show the contrast between them in the manner they express and deal with emotional distress. What about the other groups in our study? Space does not allow as detailed an account, but some general observations are in order. In the case of Singaporeans, sense of loss is less pronounced, possibly due to the relatively close geographical proximity between parents and migrant children and the more frequent visits. The relatively low-key approach to mutual emotional and moral support, also evident in the case of New Zealanders and their migrant offspring, may well be due to the fact that most of the New Zealand migrants we interviewed are male. It is a general finding across the migrant samples that men express less need for, and provide less emotional support than do women. However, there are exceptions, particularly in families where the only son migrates, as is evident in the following quote from an Italian migrant:

> *Like in my case, the eldest son is the only male in the family and so he gives a sense of security, even psychologically. And when this person suddenly disappears [migrates], the women feel a bit betrayed. Not because*

> *the women aren't capable of surviving ... They can also take it as a sign of indifference ... Especially in the south of Italy where the family is very united, very strong ... My situation as a migrant ... actually affected my cousins and aunts and uncles... cousins are like brothers, aunts and uncles are like parents.*[34]

It is possibly the refugees who suffer most because there is less outlet for their emotional needs, due to the limited amount of contact they have with their parents and other close relatives. Refugees often have very little information about their family members, and cannot assess whether they are safe. Coupled with that is a powerful sense of loss because of not being able to live close to one's family. This often leads to a sense of depression and other signs of mental distress. An Iraqi man, whose wife and fourteen children were waiting in limbo in Iran, said, 'I will get mad if I don't see them soon'.[35] An Afghan woman expressed her extreme homesickness by saying, 'when I lived in Afghanistan, we always had nice weather ... we had everything there ... here we have mental illness'.[36] And an Iraqi woman was desperate to visit her mother, because 'my mother is very sad. She cries whenever I talk to her over the phone. When I talked to her last time, I promised to go back and stay with her for a long time'.[37]

The only avenue to relieve anxiety about the safety and wellbeing of relatives in transit countries is to make regular phone calls. However, as we will discuss in more detail in Chapter 5, some relatives have no telephone, and the cost of phone calls to transit countries (mainly Iran, but also Pakistan and India) is exorbitant. In quite a few cases this has led to the decision to limit or stop such phone calls altogether. The impact on refugees of not being able to undertake or receive visits easily is further explored in Chapter 6. There is, then, for these people very little opportunity to give or receive emotional and moral support.

Without any doubt, the majority of parents and migrant children remain in close communication with each other in order to give and receive emotional support with everyday life as well as the special problems thrown up by family crises. They had been in close contact with each other before migration took place, and continue this practice. Transnational migration, however, elicits special emotional problems that are not experienced by family members who live in close proximity: homesickness, loneliness and a sense of loss and anguish during long periods of absence between parents, their migrant children *and* their grandchildren. In some cases (but not all) daughters are missed more than sons, and mothers are more important for the emotional wellbeing of their daughters than fathers. It is women who in the majority of cases do the actual 'kinwork' (di Leonardo, 1987) of giving emotional support to deal with such absence and loss. Then there are for all migrants, daughters and sons, special anxieties when parents become very frail. They face the quandary of having to decide when they should make return visits and wondering whether they will be able to see their parents again before they die.

As mentioned, the extent to which the emotional support needed by parents and transnational migrants can be expressed, and the form it takes varies both within and between samples. Many of the Dutch stoically hide their emotions in order not to upset their distant family members. They thus appear less able to express their need for emotional support than the Italians and Irish, and this means that they need to guess whether they are loved and missed rather than know this for certain. On the other hand, the Irish and the Italians tend to more readily articulate their feelings of loss and need, to the extent that many migrants from these groups often described an overwhelming sense of guilt about being far away. Such differences raise the intriguing question of what is a more satisfactory expression of emotional and moral support: expressive emotions or stoicism and restraint? Whatever the answer to this question, we are probably dealing here with differences in cultural practices and broad values of familialism versus individualism, or as described in Chapter 2, more 'communally-oriented' or more 'individually-oriented' families. These values and practices will have an impact on both the experience and construction of emotional wellbeing within transnational families.

Accommodation

When Finch (1989, p. 22) included shared accommodation in her study of family exchange, she had in mind the likelihood that family members might share accommodation over a period of time, even possibly on a permanent basis. She found that such sharing occurs during particular phases in the family lifecycle. For example, adults may live with their parents, they may return to their parents' home after separation and divorce, and older people may move in with their adult children when they are in need of caregiving. The provision of accommodation is a feature of intergenerational exchanges described by Finch (1989, p. 22) as something members of kin groups do not particularly like doing. At first glance, the topic of shared accommodation may not appear relevant to transnational family members. However, we have chosen to include questions about accommodation in our interviews in order to draw attention to the *extent* of shared accommodation during transnational visits, and to point out that in some families such sharing can be for lengthy periods and be seen as desirable by all concerned.

As shared accommodation is generally part of the broader issue of transnational visits, we will postpone an in-depth discussion until Chapter 6. A few general observations are in order, though, at this point. Firstly, in the case of *refugees*, return visits to their homelands are generally out of the question, and even visits to the transit countries where their parents have taken up temporary residence occur sporadically. Visits from parents and other relatives to refugees in Australia are almost impossible due to Australia's rigid visa regulations. For parents and other close family members who gain visas to Australia, shared accommodation on a long-term basis would be expected. Secondly, of the various migrant groups, the Irish and the Italians make the lengthiest *return visits* and usually spend this time in the 'family home', or sometimes with siblings living close to their parents. Return visits by New Zealand and Singaporean migrants tend to be shorter, but again they usually stay in their parents' home. The Dutch do so much less often, because their parents are at an age when they have moved from the family home into small apartments. Thirdly, *parental visits* to migrant children in Australia commonly involve accommodation in their children's homes and this can be of fairly long duration.

In the context of family exchange, it is important to note that migrants as well as parents frequently comment on the costs they save by being able to stay with their relatives. Remarks such as 'you have not got the expense of looking for accommodation'[38] are common. One Irish woman summarised neatly what many others also voiced, 'In Ireland you stay in their houses and they feed you.'[39]

Practical support

In Finch's (1989, pp. 31–33; see also Finch & Mason, 1993) account, practical support tends to happen mainly between mothers and daughters and centres to a large extent on assistance with domestic duties (shopping, laundry, sewing) and childcare. We asked interviewees about a variety of practical support activities, not only between mothers and daughters. As in the case of shared accommodation we assumed that practical care would primarily take place during visits. This was indeed the case, except for some specific instances of practical care 'from a distance'.

It may be useful, with 'care from a distance' in mind, to begin with a brief comment on the *refugees*. While they have very limited opportunities for visits, they nonetheless engage in significant practical activities to assist their transnational relatives, namely the writing of numerous sponsorship letters and the organisation of a multitude of visa applications on their relatives' behalf. Such activities are often unsuccessful, and a great deal of money is needed for repeated applications. To continue applying is an absolute necessity, given the difficult circumstances under which the relatives live, as well as the refugees' fervent wish and sense of obligation to ensure a safe haven for their families and, if possible, to be reunited with them.

In the case of *migrants*, there is also practical support from a distance. This is generally support provided by parents. Activities typically include renewal of a migrant's driving license or passport in the home country, forwarding mail, looking after property and bank accounts and maintaining contact with friends on behalf of migrants. One Dutch father, for example, buys flowers on behalf of his daughter to give to her girlfriends in the Netherlands on their birthdays, sends his daughter clothes and toys for her children (upon her request) and in one instance even sent an entire pram, having disassembled it in small parts for easy mailing. Interestingly, this father indicated that he did these things for the sake of his late wife. In his words:

> *Look, she also doesn't have a mother anymore. And she loved her mother, and her mother loved them too, the children ... She has to do it on her own, over there. And I think, well, I have to take on that task. I feel that is my duty, that I do it all for my wife. That's how I see it. And I do it all, with all sorts of love and with pleasure.*[40]

During visits, fathers occasionally carry out repairs and help in the garden, but on the whole their practical care is limited. Practical care provided by mothers can be very extensive and includes babysitting (both during parents' visits, and when migrants return home), some cooking and gardening, and especially care of daughter and grandchild immediately after the birth of a new baby. Dutch, Italian, New Zealand and Singaporean mothers make special visits at the time of childbirth. In the Dutch case, this is often at the special request of their daughters. In the words of one Dutch woman:

> *My mother came on her own because I phoned her [and said] oh I don't think I can handle it anymore. I might need a Caesarean. I did not see a way out, so within two weeks she booked a ticket and she was here.*[41]

The emotional support mothers are able to offer is often just as important as the practical help. Nonetheless, some mothers do an inordinate amount of domestic work at such times. One Singaporean migrant, who had just had a baby, found her mother 'priceless, she cooked dinner, she did the housework, she did the children, she did me, she did us'.[42] Interestingly, there was no indication that Irish mothers come to help their daughters at the time of childbirth. In fact one Irish woman commented:

> *Everyone in Australia, when I was pregnant said 'is your mum coming out', but I knew she would not, because 'what do you need your mother for to have a baby', that would be mum's attitude. And I was beginning to feel like my mum should be there, very strange … in Ireland, like in our family my sister had three kids, she'd just spit them out.*[43]

Her mother did not come, but three months after the birth her daughter returned to Ireland for a long visit, when 'I had a really good rest and mum looked after the baby … it was a real treat for me'.

The above examples indicate that practical help given by parents during their visits to Australia is gendered. A similar gendered pattern occurs when migrants visit back home, although male migrants tend to engage in more practical support during return visits than their fathers would when visiting Australia, particularly in the case of old–old parents. Returning sons carry out repairs, mow the lawn and do other outdoor work, while daughters help with cooking and gardening. Some take their mothers shopping or on social outings.[44] Several migrant women made special trips home to help their parents move house or help decide on aged-care arrangements.

We have noted some differences between samples as to the extent of practical care given (for example, the absence of Irish mothers at childbirth). At the same time there are probably just as many intra-sample differences in this regard. Some parents who visit (from whatever country group) are much more helpful than others, and some migrants who return home to visit do very little, while others use the visit to provide extensive practical support. The *necessity* for practical support of parents by their children is of course to a considerable extent determined by parents' state of health, as is the practical support parents are able to give their children. Another important factor is the availability of local siblings. For example, many Irish migrants come from very large families. This means that their parents often have plenty of help available on a day-to-day basis, and are not dependent on their migrant children to come and give them a hand with practical, everyday chores. In the New Zealand case, the majority of parents had a local daughter who provides daily practical help, again obviating the need for this type of support from their migrant sons and daughters.

It is important to note that emotional and practical support often go hand in hand, and that for some people to give practical support is an expression of their emotions. A New Zealand migrant gave a good example of this, when he told the interviewer that when his mother-in-law died in New Zealand, his own mother gave a great deal of practical support to him and his wife at the time of the funeral. She did this, he said, because 'just being a shoulder to cry on was not enough, but making dinner or driving us somewhere, my mum considered supportive'.[45] Practical care also closely relates to personal care and it is to this aspect of transnational caregiving that we now turn.

Personal care

In Chapter 3 we reviewed aged-care provisions in the homelands of each of the transnational kin groups included in our study. What aged-care provisions are available – and preferred – determines to a considerable extent what contribution to their parents' personal care is expected of transnational migrants. Personal care, 'nursing someone who cannot fully look after themselves' (Finch, 1989, p. 26) requires caregivers who live in close proximity. Thus, transnational migrants who need to contribute to personal care can only do so during return visits.

Several *refugees* had elderly parents or disabled siblings in transit countries in need of care. They also said that public aged-care facilities in these countries were dismal, and only family members provided personal care. Refugees are simply not in a position to make any hands-on contribution to personal care. The only help they are able to give is financial assistance to the family members who do the actual caregiving, while at the same time making every effort to bring their elderly parents and other needy relatives out to Australia so they can take care of them themselves. Such issues are generally the responsibility of sons in Iraqi and Afghan cultures. Several refugee sons in our sample were trying to bring out their elderly mothers. As one said about his mother: 'she always wants to live with me as I am her eldest son'.[46] This would only be possible if there were vast amounts of money for sponsorship, and even then, their chances of gaining a visa for an elderly parent or other disabled relative would be extremely limited.

In the case of Singaporeans, it is also the responsibility of sons (especially the eldest) to provide personal care for ageing parents. A statement made by a Singapore-based younger brother of a male migrant possibly reflects the views of the majority of Singaporean adult children on this issue, 'we are Chinese; we are supposed to stay with our parents. And especially when you are the eldest, when your parents are old, they need to stay with you and you need to take care of them for the rest of your life'.[47] This means that many parents, especially those who are widowed, live with one of their local children (preferably a son) and would expect to be taken care of by them and their wives. The knowledge that parents are secure in this respect is reassuring to migrants. As one migrant woman commented, 'my eldest brother is in the UK, so the burden of it goes to my other brother whom my parents are staying with. I am glad for that, because I don't feel so worried about that as if they were staying alone'.[48]

At the same time, many are fully prepared for the possibility that their parents would wish to come and stay with them in Australia, or that they might have to return to Singapore to assist with personal care. One woman said, 'I think, if I have to, I'll just have to take six months unpaid leave and come back [from Singapore] after six months'.[49] It should be noted, though, that at the time of interview most parents were still in good health and not in need of personal care. When parents looked into the future, some said they did not want to live with their children, so as 'not to be a burden'. They thought that if they needed personal care, they would hire a maid as paid caregiver. A widowed mother said:

> *If I grow old it is better if I don't live with them, because you don't know how you behave when you grow old. Since I have the means I will ask them to get somebody to help me in the house, and I prefer to stay on my own. And my children will keep an eye on me.*[50]

Even when elderly parents live with one of their children it is likely that those who have the financial means will hire a maid to do the hands-on caring. Local children, then, give companionship, but not personal care, and the same would probably apply in the case of migrant children who come to visit.

In many respects the circumstances of New Zealand parents are similar to those of Singaporeans: most are still in good health, and although they do not live with their local children (as their Singaporean counterparts do), they also rely on their local children for support. There is one *major* difference. The local child that most New Zealand parents rely on is a daughter, not a son, as in the Singaporean case.[51] Most New Zealand migrants happened to be sons, and given the cultural expectations of Western countries that women, rather than men, are caregivers, there was no great expectation that they would make a contribution to their parents' personal care.

In the case of migrants who live further away, much depends on the availability of aged-care provisions. The Italian state provides few aged-care facilities. In the case of the Irish, the state provides more, but in both cases there is a great deal of resistance to any form of institutional care. Most Italian parents preferred to be cared for in the home of one of their daughters, when they were no longer able to live independently. One male migrant said:

> *In Italy, it's the children who have to look after their parents ... it will be hard when mum gets old. She will probably go to my youngest sister, because it is usually the youngest one who looks after the parent ... an old people's home would be a disgrace for the family.*[52]

This situation, in the case of the Italians, may demand a contribution to personal care from migrants, either to share the burden of care with local siblings or to give them respite. In fact, we found it was very common for the Italian-born migrant daughters to make return visits for this reason. We return to this issue in Chapter 6. Like the Singaporeans, if children were not able to provide personal care, then the Italian parents preferred to be cared for by a live-in female carer or housekeeper.

Irish parents, on the other hand, expected to remain in their own home, cared for when necessary by one or more of their local children who lived on the same land, or close by. Given the large size of Irish families, there was less need for migrant children to assist with personal care. Thus, one Irish mother, speaking about her local children, expressed confidence that they would all look after her. She added that her sons often said: 'if there is anything you need, just phone immediately, and we'll be there in a tick'.[53] Migrant children are expected to visit, 'to spend time with mother', but in all but one case not to provide personal care. Of course in the always hoped-for situation that they would repatriate, migrants might have to take their share of caregiving like all other local siblings.

Dutch parents have a wide range of aged-care provisions available. Those in our sample prefer to live independently and when personal care becomes necessary, they will enter institutional care rather than live with their children. In that situation, local children still provide a great deal of emotional and practical support, but personal care tends to be in the hands of professionals. Migrants' contribution consists of return visits to provide emotional support (as in the case of the Irish) but not personal care. Thus, of the three migrant groups that live far away from their relatives, only the Italians were likely to make return visits specifically to provide personal care.

Factors that impact on the capacity and obligation to care

In Chapter 1 we suggested that the capacity, sense of obligation for transnational caregiving and negotiated commitments are affected by macro-structural-, meso- and micro-factors. As macro-structural factors relate directly to the policy implications of transnational caregiving, we postpone discussion of these until Chapter 7, although many are also dealt with in Chapter 6 because they affect the capacity for transnational visits. Meso-factors, we defined as issues to do with community expectations and obligations. We incorporate these in our discussion of micro-factors – the main focus in the next section.

When Finch and Mason (1993, pp. 10–14; Finch, 1989, p. 13) asked in their research on family obligations 'who does what for whom', they answered this question by pointing to a combination of *normative guidelines* and *negotiated commitments* between family members. We concur with such findings: when family members carry out caregiving transnationally, both normative guidelines *and* negotiated commitments apply. Central to the normative guidelines under which family members accomplish transnational caregiving are general cultural expectations regarding the caregiving responsibilities adult children have towards their parents, as well as responsibilities due to gender, marital status and birth order. Such normative guidelines, however, may be circumvented, ignored or rejected because of specific negotiated commitments between family members. These, in turn, depend on personal biographies and shared family history. Let us illustrate.

When asked, most migrants, regardless of country of birth, indicate they feel a sense of moral responsibility to look after their parents as they age. They think that such a moral responsibility remains even for those who live very far away, and in a number of cases they see such caregiving as a repayment for the fact their parents looked after them when they were young. Some also invoke communal codes of honour and obligation that they need to adhere to. In the case of Afghan refugees this was the notion of Qaum kinship obligation. Singaporeans would preface sentences with 'coming from an Indian family',[54] or 'we are Chinese and thus ... '[55] or 'this is a must in Islam'.[56] Many Singaporeans also make specific references to 'filial piety'. Italians talk about the role of the family and the Italian expectation that 'good children' care for their parents. Several Dutch respondents suggest that there are regional differences, people from Friesland, for example, being seen as more emotionally reserved then people from Brabant to the south. Such cultural ideas often incorporate obligations based on gender, marital status and birth order.

We turn first to the case of the Singaporeans. Contrary to what is often assumed to be the gender hierarchy of care in Western societies – with women carrying the burden of responsibility – among Singaporeans it is mainly the oldest son who carries the prime responsibility for filial piety (although his wife may do most of the actual care work).[57] One Chinese–Singaporean migrant, who had an older sister living in Singapore, said, 'as eldest son, you tend to play a more important role in terms of taking care of the parents'.[58] If the eldest son lives abroad and is therefore not able to provide a home to elderly parents, a younger son has to shoulder the responsibility. To live abroad, however, does not absolve the eldest son from the responsibility of sending financial support to his parents. This was illustrated nicely by the words of one Singapore-based younger son who lives with his mother and looks after her financially, speaking about his older migrant brother who did not provide financial support, 'yes, filial piety, caring for one's parents or elders is a must in Islam. Without a doubt. It is in fact an obligation. My eldest brother does not fulfill his obligation'.[59] The fact that sons are seen to have the major responsibility lets daughters, especially married ones, 'off the hook' to some extent. One female migrant commented, 'they would not ask for help [with finance]. I think it is to do with pride and the fact that they are parents, and I have been married off. So for a lot of the caregiving at the moment, they rely on my brother. He sends them money and that sort of thing'.[60] Having said this, though, she added 'I will probably contribute when my mother retires'. Another woman said, 'coming from an Indian family, they really don't like to accept money from daughters ... I suppose they will if it is a token amount'.[61]

In the case of Iraqi and Afghan refugees, the eldest son also carries a special responsibility towards his parents. A corollary of this is that he often will not have the capacity to take responsibility for the care of his in-laws and will not send financial support to his wife's parents. This can create conflicts between husbands and wives. One Afghan woman described such a situation, 'my husband does not send my parents money. He is saying we have enough expenses in Australia'.[62] This woman gave presents, including some money, to our interviewer to take to Iran to give to her relatives. However, she did this secretly, so her husband could not see what she was doing. An Iraqi woman complained that her husband:

> *Calls his family more than me. Whenever I want to call my mum, he says, 'it costs a lot'. But whenever he wants to call his parents he does not bother and he may talk for half an hour or more without thinking of the costs. Once I called without telling him. When he saw the telephone bill he noticed it immediately ... I tricked him and convinced him that he knew about it.*[63]

Such findings concur with Pahl's research (Pahl, 1989, 2001a) in the United Kingdom, who found that husbands frequently have control over any financial decisions taken in their immediate family. In the case of the refugees in our study, the majority of husbands controlled decisions concerning financial support to be given to their in-laws. There are numerous examples of this, even in cases where in-laws actually did receive financial support. An Iraqi woman said that her husband paid for all her travel expenses to Iran to see her daughter and her mother. He even insisted on her purchasing respectable gifts for all her family. She suggested that he had to protect his honour and therefore spend generously on his wife's family. This woman then added, 'some Iraqi men just help their own parents. Among some Iraqi there is always a fight between husband and wife, because the husband sends all his money to his own family. My husband is not like them'.[64]

Interestingly, if the wife has money of her own, such a problem will not arise. In other words, her husband does not exercise control over her money. Also, in some cases, a degree of negotiation between husband and wife is possible. An Iraqi woman showed through her comments that she felt she had some power to control her husband's decisions regarding financial support *even* to his own family. Her husband's relatives lived as refugees in Sweden. When her sister-in-law called from Sweden to ask for money, she said to her husband, 'if you send them money, I [will] get very angry and move myself and my children out of your house'.[65] She had then told her sister-in-law on the phone, 'what do you think we are doing in Australia? My husband has no job and we have not enough money'.

The above examples draw attention to the possibility that the desire of one marriage partner to give care to his/her parents may be thwarted by his/her spouse. In Western literature (see, for example, Pahl, 1995; Singh, 1997) there has been considerable interest in the potential conflicts between marriage partners over financial issues, but there has been no research on the effect of such negotiations between husbands and wives on the *caregiving* capacity of migrants. Our research reveals that transnational caregiving responsibilities can indeed generate conflicts between refugee couples and that some refugee women are not allowed by their husbands to give their parents financial support. In fact, a local social worker who deals with migrant issues indicated to us that transnational family obligations were a major cause of conflict in many refugee and migrant families, and a frequent causal factor for divorce.

Male power over household finance is also an issue among Singaporean migrants, but to a much lesser degree. It showed in two ways. First, it was clearly the father (rather than the parents together) who had decided what financial support would be granted to children during the time they attended university in Australia. Second, several female migrants hinted at the need for negotiation with their husbands about finances, as was evident in the quote of a Singaporean woman at the opening of this chapter. She warned her non-Singaporean husband at the time of her marriage that although he did not need to contribute, she had responsibilities to her extended family and intended to keep these.[66] Most Singaporean women had been or were in paid work at the time we interviewed them, and had already adopted a much more egalitarian approach to their marriages than would have been the case with their parents. One migrant woman of Indian–Singaporean descent said that her parents made their son-in-law promise that he would 'send me back at least once or twice ... because in Indian thinking, the husband has to give his wife permission to visit her side of the family'.[67] She said this tongue-in-cheek; it was clearly not a set of values she believed in herself.

Birth order and marital status appear to be of much less significance in the other cultural groups, but the issue of gender and power remain pertinent. One Dutch migrant wanted to go and visit her father on his 80th birthday, but her husband exercised subtle pressure on her not to go. She said, 'I want to go and see him. ... I don't know if I can. My husband says: don't do it, but I feel a bit guilty if I don't'.[68] In the end she did not go. In some other instances (of Dutch migrants married to Australians), husbands do not provide much help in making parents welcome when they visit.[69] But these are minority cases, and concern emotional support rather than financial help. Gender, on the other hand, also plays an important part in transnational caregiving among the Irish, Dutch, Italians and New Zealanders. The difference is that *daughters* rather than *sons* are seen as primary carers. When parents have daughters in close proximity, as was the case with many Irish, New Zealand and Italian parents (and a few of the Dutch), male migrants tend to be relieved of any responsibility. The opposite does not quite apply: migrant daughters continue in most instances to feel responsible and guilty, even if parents have local daughters who can give care.

This means that a daughter who is abroad may become defined as a 'bad daughter', whereas a son may be seen as a 'good son' – regardless of the amount of caregiving and support each has accomplished. An Irish migrant, for example, in spite of the fact that he did not provide personal care to his elderly mother, was perceived as a 'good son' because he maintained close contact with his sisters in Ireland providing them with ongoing emotional support. Both gender and distance played a role in releasing him from the obligation to provide more care. He felt, in fact that he was the lucky one in this situation, saying 'you get away with a lot when you're not living with your elderly parents ... you don't have to do anything, really, I mean just call on the phone'.[70] A New Zealand migrant was similarly able to be a 'good brother' by giving his local sister emotional support. She was burdened with extensive responsibility for her parents and her in-laws, and was under significant stress. Her Australian-based brother counselled her throughout. One respondent gave a sibling perspective on such situations when she said about her migrant brother 'he has that choice to decide how much time he gives to his parents. I don't. So he is in a controlling situation'.[71] She then added that this means her brother 'can be the good child and maybe this is why he is looked upon so'.

Such examples are in profound contrast to the experiences of a Dutch woman who lacked the capacity to return back home to support her mother when she was dying of cancer, for reasons of finance and responsibilities to her own immediate family. Her sister and other relatives in the Netherlands expected her to return and when she did not, this generated a family stigma against her as someone who was not a good enough daughter. This was aggravated by the fact that she had a cousin in Australia who, due to her work, had been able to make numerous return visits to her own parents, and was therefore defined as a 'good daughter'. Family pressures and her own sense of unresolved guilt preyed on 'the bad daughter' for many years after her mother's death.

Legitimate excuses not to care from a distance

Do migrants and their transnational families use long distance as a legitimate excuse not to provide care or to provide less care than local siblings? Some do, and this occurs across all samples. There are instances where the fact that migrants continue to visit and communicate is seen as a sign of being a 'good child'. There may even be a tendency to single out migrants as especially good for maintaining contact. An Irish migrant sums this up in a lengthy comment:

> *I am the one who has moved away, and they all seem to be immensely understanding of that. And they seem to appreciate the fact that I come the furthest to see everybody. Like I'm the one who has always gone back, I've taken children on the long flight to go back and see the family, and I've spent the money and taken the time to do all that, and they have done what they can at their end, to make my trip easier, you know, meeting flights, waiting around while flights have been delayed, given me accommodation, and making sure I make the flights back out even at very unsociable hours. They have never complained and they have accommodated me in every way.*[72]

It may also be suggested that when migrants indicate in interview they have siblings back home to provide care they in fact invoke the excuse of distance. They do not see it as necessary to jump into the breach because they are so far away. One migrant showed this in her comment: 'I have never felt very guilty about not being there, and maybe actually did not think there was a place for me there'.[73] She is the only daughter, and her local brothers act as caregivers for their widowed mother. Such instances, however, are rare. Much more common are situations where migrants express guilt or local children indicate resentment against their migrant siblings for not providing sufficient help with the care of their parents. The excuse of distance clearly is not seen as legitimate in those cases.

In fact, the relationships between local and migrant siblings appear as key factors in transnational caregiving. Migrants are often dependent on the goodwill of their siblings for information, and local siblings may exclude or manipulate access of migrant children to their parents. Local siblings may be at a special advantage at times when parents begin to suffer from memory loss and no longer recognise or acknowledge their migrant children when they visit. There are a number of specific examples of such situations, which we will return to in Chapters 5 and 6. On the other hand, the local sibling may be especially burdened in a way that the migrant never experiences. It is not surprising that a sense of guilt may occur on the part of migrants, and resentment on the part of local siblings. We suggest that the extent to which guilt and resentment are present is due to the degree of equity and reciprocity, which has been accomplished between transnational family members.

At the same time, there are unique issues related to personal and family history, which have little to do with long distance. The aggression shown by a father to his migrant daughter, when she came to visit him close to the end of his life, although probably due to dementia, was in effect the endnote on a long history of hostility and authoritarianism started well before she even migrated. A migrant felt that her parents had never supported her financially or emotionally after her emigration to Australia. In fact her parents had given her substantial financial support; that she chose not to remember this can be explained by her anger over her parents' negative responses to her sexual preferences (something unrelated to her living at a distance). We have already mentioned the case of a migrant who fell short of his Singapore-based brother's notion of a 'good son', because he did not support his parents financially. Interestingly, a relative provided a legitimate excuse for him not giving economic support by invoking an aspect of his personal history: 'he had not really been brought up by his biological mother'.[74] In other instances where we found that migrants went 'against the grain', in other words, did not provide support where cultural obligations, gender marital status and birth order presumed they should, we do not have sufficient information to be able to assess the full range of motivating factors. But where we do have the relevant data, it is evident that past family histories do inform intergenerational caregiving across borders to the same extent as they might when parents and adult children live in close proximity.

Conclusion

Reciprocity in intergenerational relations has been a central theme in this chapter. In a sense, this is not unexpected. Most contemporary sociologists and historians who have studied intergenerational caregiving have found 'the continued centrality of reciprocity to human relationships' (Thane, 1998, p. 206), with parents giving to as well as receiving from their adult children.

Possibly not quite so expected is the fact that reciprocal relations between parents and adult children carry on across national borders, and even when family members live far away from each other (but see Stafford, 2005). As we found, parents continue to support their migrant children financially, when they have the capacity to do so, and their emotional care and support remains central throughout their lives. In turn, transnational migrants – like their local and translocal siblings – repay parents for care given before. They do this financially where required or culturally appropriate, and emotionally until their parents die.

Possibly unique to transnational caregiving is the degree of exchangeability we observed between types of support and care. This is especially relevant to the provision of personal care, a type of support that in most cases transnational migrants cannot contribute to. When Italian migrants decide to 'forego' their inheritance, thereby promising future financial benefits to their local siblings who are providing personal care to their parents, what these transnational migrants are in fact doing is substituting economic support (which is in their power to give) for personal care, which they cannot provide. When refugees send money to their kin in Iran so that they have the resources to give personal care to elderly and incapacitated relatives, they effectively do the same, substituting financial support for personal care. Through such substitutions, then, reciprocity between the generations can be maintained, even from a distance.

5
Communicating Across Borders

> *Well, in the first place, phoning was very expensive, so, we didn't phone very much. We would write, and when I think about that now, I don't understand how I've ever been able to do it. I had a notebook, I still have it somewhere, and so I would write down to whom I had been writing, otherwise I couldn't remember, and ... I wouldn't put a letter away before I had answered it first. ... In that first year I wrote, I think, about 370 letters. That's more than one letter a day. Handwritten. In those days I didn't like to use a typewriter, I didn't have a computer then. ... I kept that up for a couple of years, but my neck started to hurt from all that writing. Everybody else would write to just one person, but I had to write back to everyone. My husband didn't like writing very much, and I like it, I mean, it wasn't that I hated doing it. So, then I changed to the typewriter, and now it's all email, of course, that's ideal. So yes, that was the only way, the most important means of contact, by mail.*
>
> (Dutch migrant daughter)

The recent and dramatic revolution in telecommunication technologies has transformed the ways in which people interact across distance. Many of the parents in our study are able to remember a time when it took several months for news and people to travel from one side of the world to the other. Now, entire family networks regularly and instantaneously share information across that same distance. This 'death of distance' (Cairncross, 1997) has important implications for transnational caregiving within family networks.

We begin this chapter by documenting the 'miracle' of transnational communication technologies from the perspective of kin who *want* to stay in touch across distance and thus rapidly adopt new technologies in their efforts to maintain family networks. We then assess the extent to which the adoption of new technologies appears to enable more multifaceted experiences of family life across distance and borders. However, we are very aware that 'happy families' communicating across distance are not the full story. Many people still lack easy access to communication technologies, rendering some families more 'global' or 'transnational' than others (see also Wilding, 2006). This is a particular issue for refugee families living in transit, and also for those with ageing bodies and minds that are ignored by creators of new technologies.[1] For others, it is actually the very *capacity* to communicate that creates the problems as long-term family disputes or the need to live up to cultural obligations such as 'the good daughter' or 'the good son' are no longer muted by distance. Rather, they, too, are regularly and instantaneously transported across the miles, informing the negotiated commitments at the basis of all family relationships. Thus, we demonstrate in this chapter that the simple increased availability of communication technologies is only part of the story of transnational caregiving. Of special interest to us here is the ways in which families incorporate such technologies in their ongoing negotiation of relationships, expectations and obligations of transnational care and support.

Most of the participants in our study were thrilled about the recent advances in communications technology, making comments such as 'communication is getting better, distance is getting smaller; *anything* is possible'.[2] Yet, interestingly, new technologies do not automatically replace older, less useful forms of communication. Rather, email, webcam, mobile phone text messages, videos, faxes and websites are used to supplement existing modes, such as letters, gifts, audio tapes and photographs. Indeed, recent scholarly attention to the implications of new technologies for social relationships risks overlooking a very strong point of continuity: new technologies are incorporated into long histories of families using all available means to keep in touch with their kin overseas (see also Miller & Slater, 2000; Dimaggio et al., 2001; Wilson & Peterson, 2002; Benton, 2003). That is, communication technologies are important in enabling people to do what Michaela di Leonardo (1987) has termed 'kinwork'.

For di Leonardo (1987, p. 443), kinwork is best defined as 'the conception, maintenance, and ritual celebration of cross-household kin ties', which is accomplished through such acts as making visits, sending letters, exchanging telephone calls, organising holiday celebrations, keeping and sharing family albums and sending gifts and cards. Among the Italian households that di Leonardo studied, kinwork was done primarily by women, who were usually the ones who made the time and developed the skills necessary for 'maintaining these contacts, this sense of family' across time and space (di Leonardo, 1987, p. 443). In trying to understand *why* women do this work, di Leonardo points out that an investment in kinwork is an investment in future obligations to provide care and support, particularly between mothers and daughters. Thus, every letter, gift and card that is sent and received and every telephone call that is made acts as a reminder of the mutually supportive relationship inherent in the kin connection, and of the expectations and obligations of care and support that are implied in that relationship. In Finch and Mason's terms (1993), each act of kinwork is a part of building the history of negotiated commitments within the family network.

Di Leonardo (1987, p. 452) fears that 'this investment was in most cases tragically doomed' due to the ongoing decline of intergenerational obligations, but our study provides clear evidence that kinwork continues to occur, and that it continues to be important in demonstrating existing as well as establishing future expectations of care and support. Moreover, this is in spite of large-scale migration processes dispersing families across national borders. What has changed, in large part due to technological transformations, are the tools that people use to communicate. A brief history of the ways in which families have responded to the shift from letters to telephone calls, and then to faxes and emails, helps to demonstrate this point.

Historical shifts in communication

For those who migrated prior to the 1990s, the most common means of keeping in touch was at first by letter. As the quote at the opening of this chapter shows, one migrant, a Dutch woman, commented that after her migration to Australia, 'in the first year I wrote, I think, about 370 letters'.[3] Parents, too, reflected on the extensive letter writing of the past, making comments such as 'in the beginning we did a lot of writing'.[4] These letters were usually sent on an exchange basis – when a letter was received, there was a sense of duty to send one back as soon as possible. Because not everyone replied immediately, and letters travelled at different speeds to and from different destinations, families had quite varied experiences of frequency in their communication. The most frequent rate of exchange was once a week, but many exchanged letters on a more occasional basis. As one mother in Ireland described:

> *The letters were good, because they had to put everything into the letter at the time, and [my daughter] was a very good letter writer, she was very descriptive. But they weren't frequent, that was the only thing, they were always fabulous letters, but I mean you could be waiting for six months for one!*[5]

The letters were often accompanied by photographs and audiotape recordings of songs, messages and 'the first words of the children'.[6] This practice of sending audio recordings was so widespread that it appears there was obviously something very attractive about being able to 'hear' geographically distant kin even before the telephone was affordable and readily available. For example, one migrant talked of how grateful she was to her sister in Italy, who supported her during her divorce by sending audiotapes full of reassuring messages. In a similarly strong testament to the significance of the voice, an Irish migrant talked of how her mother's tape recordings of her singing old songs in the kitchen would transport her momentarily back into the comfort of the family home. It is clear that the immediacy of the voice in both recordings and on the telephone is a very powerful experience of connection, which often serves to erase the sense of time and distance. For example, one Dutch migrant talked of how her mother's voice on the telephone gave her the impression her mother remained forever young; although she knew from her visits that her mother's body and face had aged, her disembodied voice had remained unchanged throughout that time.

Even for those who wrote infrequently, letters remained the most efficient and effective way of keeping in touch until very recently. Indeed, they remain so for the refugee families today, for reasons we discuss later in this chapter. A woman who left Ireland in the 1960s was typical in saying:

> *I would write regularly; there was probably not a week that we went through that I didn't write a letter ... We didn't do a lot of phone calls, it was very expensive in those days, sort of Christmas and Easter perhaps, or birthdays.*[7]

The telephone was used only for emergencies or on special occasions. In comparison with letters, telephone calls were difficult, expensive and generally unsatisfying. One Irish woman recalls her early calls from Australia as follows:

> *At that stage my mother didn't have a phone in the house, so I'd ring the next door neighbour, and they'd run over and get somebody, and they'd come back ... and, well the neighbour was my aunt, so it was like you'd get one of the cousins or whatever and have a bit of a chat with whoever answered the phone, while someone would hop over the wall and get somebody from our family.*[8]

Even aside from the difficulties of organising a telephone call, prior to the 1980s there were common experiences of echoes, interference and lines simply dropping out mid-conversation. Furthermore, the high cost of the calls made it hard to concentrate on the conversation while 'keeping one eye on the clock'. For example, an Italian migrant said:

> *The phone calls were not very good because they were brief. I felt like they didn't achieve much. It was: "hi, how are you?"; it was basically like a greeting more than a proper conversation.*[9]

For similar reasons, telephone calls remain problematic for the refugees today. But for the migrant families, telephone calls have since been transformed from an occasional means of communicating to the most regularly used tool of all. In large part, this is because the telephone is now relatively cheap and reliable in comparison to the letter. The cost of a telephone call from the mid-1990s onwards is compared by many migrants to the cost of a cup of coffee, and has the added advantage of providing an instant connection to home and family. In comparison, a letter requires planning, time and effort to write, a trip to purchase postage (which has become expensive relative to telephone calls) and then a wait of at least a week – but usually longer – for a reply.

In contrast with the letter, many people say that a telephone call requires no particular effort beyond having a desire 'to chat' and a telephone – now available in most households. The difference is clear in comments such as that by a Dutch woman who had migrated to Australia in the 1970s, and changed to telephone calls as soon as possible:

> *I think I'm not very good on paper – I've always done it, writing, because I got all these letters from my mother. Every week I wrote one sheet of airmail paper, but I haven't done that anymore for the last 15 years. Now, we call.*[10]

Most telephone calls are between 20 minutes and one hour in length, and most people make telephone calls on a weekly basis – almost always on a Sunday. The pattern of the weekly Sunday call is probably the outcome of convenience. Sundays are not usually interrupted by work or school commitments, and with the global time differences – between six and eight hours for Europe and Western Australia – they present the largest window of opportunity for catching both migrant and parent at home at an appropriate time simultaneously. The impact of time zones also explains why the Singaporeans are the only group who did not follow this 'Sunday call' pattern. Rather, both parents and migrants said that they make calls to each other 'all the time', something made easier by the fact that Western Australia and Singapore share the same time zone. For the other groups, a Dutch man was typical in his statement that 'we made a point that every Sunday morning [in the Netherlands, but afternoon/evening in Australia] I would phone my parents'.[11] The main exception is the 'special occasion call', when kin are contacted on their birthdays or other commemorative occasions such as anniversaries and Christmas Day.

Unlike the pattern of reciprocal exchange for letters, telephone calls are often considered the responsibility of the migrant alone. The main exception is again the case of Singapore, a sample in which both parents and migrants make frequent calls without apparent regard either to patterns of exchange or to one person taking responsibility. While other sample groups also include some exceptions, the most common pattern is of migrants making the majority of the telephone calls. This expectation is usually justified on the grounds that call costs are less expensive in Australia than elsewhere around the world – except for Singapore, which we were told has a similar fee structure to Australia. The shared time zone and low call costs also seem to enable the Singaporean migrants to maintain regular telephone contact with their friendship networks as well as their families, something that was much less evident with the other migrant groups.

Migrants commented on the competitive deals they are able to get from particular telephone companies in Australia. For example, an Italian woman said:

> *After we joined [this telecommunications company], we can call Italy any time of the day, seven days a week, for 29 cents a minute. That's brilliant ... because, still now [year], all the other companies they charge you a dollar per minute. So after we joined this company, we really – things really changed. Now, any time we think to call, it is no problem at all.*[12]

The same woman contrasted this situation with that of her family in Italy at the time:

> *They still ring every two weeks, but if they ring we say "just hang up and I will call you back". But they are getting better in Italy too now, the competition ... I think it was three dollars a minute to call from Italy, and I think it is a bit cheaper now ... [a relative] rang four times and had to pay 85 dollars, and he was talking only five minutes, probably. Or when he used a public phone, ten dollars would run out in two minutes.*

However, call costs are not the only factor in the decisions about who calls and who is called. For some people, the emphasis on the migrant doing the calling is simply a matter of practicality – the migrants tend to have busy work lives or young children to care for, and so are seen as being less available to be called. Most parents, on the other hand, have relatively quiet routines, which means they are often available in the mornings and evenings, when calls are often made. In other cases, the migrant feels it is their responsibility to call because they are the ones who left, thereby creating the need for telephone calls in order to keep in touch. Many parents, too, mentioned that this obligation to perform the kinwork of telephone calls follows automatically from the decision of their child to migrate.

While most families now use the telephone as their most common means of keeping in touch, this is not to say that letters have disappeared altogether. Rather, letters have become useful for rare and special occasion communications. Many families use letters to inform each other about illness, death, divorce and other potentially sensitive information. For example, one family talked of how distant grandparents were informed of their grandchild's mental disability in a long and carefully thought-out letter, full of reassurances, rather than in their regular telephone conversations. Letters are also sometimes sent at Christmas time – usually in the form of very lengthy photocopied 'newsletters' that provide an annual report on events, accompanied by photographs.

Letters are also sometimes sent because they are perceived as more tangible and reliable expressions of support and care. This is clear in the case of one Irish woman, who became concerned about her mother after a particularly bad telephone call. Her mother was worried that she might never see her daughter again, and would be left to die alone. In a long letter written immediately after the telephone call, the daughter declared her promise to return at a moment's notice if needed, and gave a detailed description of how this would happen. Years later, this letter was remembered as a positive turning point in their relationship. It was carefully stored away, able to be read over and over again in times of doubt or difficulty.

The special 'tangible' quality of the letter is also reflected in the fact that sometimes it is the letter as an object, rather than its content, that is most important and memorable. Many people talked of their excitement at finding a 'blue airmail letter' in their mailbox, particularly when they were battling the initial sadness and homesickness immediately following migration, often acknowledged as a special time in the family history. The tangibility of the postcard apparently serves a similar function: one Dutch migrant sent a postcard to his mother on a weekly basis and, while her dementia meant that she often forgot the content of the message, the postcard acted as a constant and visible reminder of her son's care and love.

While letters are still reserved for some communications, there is nevertheless a clear pattern in which a rapid decline in costs for telephone calls throughout the 1990s resulted in a dramatic increase in calls and decrease in letters. But there was also another transition at work: during the 1990s, many migrants began to reduce the number of regular telephone calls and letters and increase their communications by email and fax. For example, a Singaporean woman who first came to Australia in the early 1990s said:

> *When I was studying, I used to write a lot actually. I used to write loads of letters back in my first and second year at uni. But by third year, I was struggling to get something out. Same goes with my friends as well. They used to write to me all the time. But now, with email, it's been great – we email each other every day!*[13]

The fact that many of the Singaporean migrants have come to Australia to complete their tertiary education might lead us to conclude that their education was the reason that they used email communication more readily (see also Anderson & Tracey, 2001). However, it is not only the Singaporeans who follow this pattern. Several interviewees in the European and New Zealand groups reported similar stories, and not all of them have tertiary education.

Emails are usually sent with great regularity. In contrast to the telephone and letter, they are exchanged several times a week and, in some cases, even several times a day. They are usually 'a few lines', with anything longer prompting a telephone call or letter. One Italian woman found that emails were enjoyable because '[they] send me funny pictures, or funny sound bites'.[14] An Italian man described the different forms that emails might take:

> *I send emails because it's something I want to do ... never serious things, never things that will prompt a serious response from them; that is in the sense that it's not every day five times a day that you will send serious emails in which the other person has to sit there and think. It is the kind like you would say with a friend at a cafe, 'hi, how are you' ... that is, light-hearted stuff. And then every so often there are a few emails in which you really ask how they are, what they think of you, but this is something less frequent. It's not that every day you will ask these things.*[15]

For many people, emails seem to be the ideal way of saying 'I'm too busy for a long chat, but I'm thinking of you.' As one Irish woman explained: 'We're all very busy, we just don't have time to get into the stuff in our minds, so we usually go "lovely to hear from you, can't talk now, but I'll" – whatever, you know.'[16]

The fax appears to function in a very similar way to email. Only a small number of the migrants from Italy, Ireland and New Zealand reported buying and using fax machines, and none of the refugees or Singaporean migrants used faxes to communicate. This is in distinct contrast to the Dutch, where nearly a third of the families own and regularly use a fax machine. Faxes are considered excellent for communicating with older parents, in particular. As one man explained:

> *When I left I gave them a fax machine, because computers were a little bit too complicated for them. With a fax it is a lot easier. You can write it like a normal letter and put it in. We made a concerted effort every week to write a fax, normally on the Sunday night, and write what happened that week in as much detail as possible, normally like two pages. There is a little bit from [my wife] in there, in English, and as much Dutch as she can put in there.*[17]

Like the letter, the tangibility of the fax is also part of what makes it desirable. Rather than receiving an electronic message, there is a 'piece of paper on the other end'.[18] People talked fondly of how, if a fax arrived when they were out, they would see it when they got home: 'sometimes it hangs [from the machine], sometimes it's on the floor'.[19] Equally heartwarming was waking up to see a fax waiting for you: 'you get up in the morning and there's a message there for you, you can read it in your own time rather than be on the phone and trying to talk to your kids'.[20]

Like a letter or printed email, the fax can be filed away and looked at again later, or pinned up as a reminder of the sender, and shared among other family members at a later gathering. Also like a letter, the personal markings of the sender are more visible in a fax; the handwriting style is remembered from the past, and new inflections are noticed in the present – perhaps as a result of a child developing its writing skills, or an older parent becoming less firm in their grip on the pen. Multiple authors of the message are more identifiable by the different handwriting styles, and a fax might include hand-drawn diagrams that support the text of the message, children's drawings or a copy of an article from a newspaper or magazine that is thought to be of interest to the recipient. In these respects, the fax is able to replicate many of the desirable elements of the letter and give greater depth and texture to the communication than is possible in email. However, the immediacy and convenience associated with the email is also a clear advantage of the fax machine: messages can be sent without the additional trip to the post office; they are received within minutes, as opposed to the days or weeks that a letter takes to travel; and the conversation can be conducted at the convenience of the participants. As one migrant from New Zealand said:

> *I prefer the fax and I know Mum prefers the fax because you scribble it down and send it, and then – sometimes you don't want to talk to them. Not that you don't want to talk to them, but you know that if you pick up the phone and start talking, you'll talk about this and that, and that and that, and ... it's just like, no. This is what I want to know, and shoot it through. And I know Mum prefers the fax because ... she doesn't have to think about whether I'm going to be there. I just get it when I get it, sort of thing.*[21]

Indeed, one common practice is for people to use the fax, mobile phone text messaging or emails to organise their telephone calls to each other. Rather than take the risk of someone being busy or not at home when a telephone call is made, an email, fax or text message is a convenient way of saying 'thinking of you; can you talk now?'.

Changing patterns of connecting

It is clear that, as new technologies of communication become available, people adopt them in order to remain connected (see also Urry, 2004).

However, new forms tend not to displace old forms of communication completely. Rather, we found that different tools are used for different purposes in different historical moments. For example, as described above, the telephone has transformed from a 'special occasion' tool of communication to a relatively routine and commonplace means of keeping in touch. In contrast, the letter has shifted from the most common to a relatively rare, 'special occasion' means of communicating. Email and faxes, on the other hand, enable new experiences of transnational communication to be constructed. They have characteristics of being simultaneously instantaneous and deferrable, which neither the telephone nor the letter were able to achieve on their own in the past. At the same time, they also continue many features of the letter and the telephone; indeed, both the fax and the email could easily be described as letters sent by telephone.

This continuity in the *fact* of communicating, in spite of shifts in *how* communication occurs, is an important finding in our study. It reinforces our point that past assumptions of increased social fragmentation are overstated, in spite of increased population mobility and individualising tendencies (see also Ortner, 1997; Urry, 2000). Although migration does arguably serve to disrupt or even momentarily fracture the family network, it is also clear that families put great effort into repairing the fractures and maintaining connections, even across multiple generations (Baldassar, 2001). The swift adoption of each new communication technology is clear evidence of these efforts.

While continuity in the desire to maintain family relations across distance is important, we also suspect that the introduction of new technologies is changing the ways in which social relationships are understood and practised, by transforming underlying notions of 'proximity and distance, closeness and farness, stasis and movement, the body and the other' (Urry, 2002, p. 271). For those who use technologies to keep in touch, we can see three key shifts in how transnational family life is understood and practised: first, in terms of who communicates; second, with the emergence of a new sense of distance and proximity; and third, in relation to the power or capacity to communicate.

Who communicates?

We mentioned earlier that the shift from letters to telephone calls usually means that a practice of exchanging letters is replaced by a pattern in which the migrant makes most telephone calls to the parent. In both cases, parents, and particularly mothers, tend to operate as the 'hub' of the family network, through which the content of letters and telephone calls are distributed. They are the central contact for all of their children and grandchildren, both local and migrant, and pass on family news around the network. But the shift from letters to telephone calls has an important consequence for the management of information in this network. While a letter might be passed around an entire family network, for each person to read for themselves, a telephone call has no such record that might be replayed on numerous occasions. Rather, it is up to the parent to interpret the news of the migrant and translate it during subsequent conversations with other kin. This has the effect of further strengthening the hub, already a highly valued and influential location, as its occupants have an even larger degree of control over the distribution of knowledge and gossip throughout the network (see also Urry, 2003).

Many migrants, in particular, will redirect their communications away from the hub when they feel that they are not receiving 'good' information. To some extent, kin are always trying to 'read between the lines' of communications to check that what they are being told is truly the case. Indeed, the telephone is in part a desirable technology because most believe that the tone of voice, pauses and silences are more revealing than the words that are spoken or, in the case of a letter or email, written. The decision to look for additional sources of information is particularly common when migrants develop concerns about the health and wellbeing of their parents. Local sisters are often considered important sources of 'hidden' or alternative perspectives, with which the migrant can test their understanding of a situation. It is generally accepted that local kin will often have a more 'true' and full picture of the situation. As one Dutch migrant reflected:

> *My mum will never say if she is not feeling well, but you will notice it afterwards and she'll say, 'Oh well, I wasn't feeling well yesterday so I went to bed a bit earlier'. And I think to a degree that has now extended to us feeling far away. So [they think], 'we don't want to bother them with that because they can't help us', to the extent that my sisters could. So I think if they really had a need that they would probably go to my sisters first. And then we would hear about it via my sisters or afterwards.*[22]

Of course, this ability to gain more complete knowledge is unequal: it is much easier for a migrant to gather a range of sources of useful information than for kin in the home country to obtain reliable information about the migrant. As an Italian migrant explained:

> *I rely on more people, on more sources. They rely only on me, and so I think it's probably been more vague for them than it is for me. Although, on my behalf, there are more people that I need to keep track of. ... For example,*

I asked my sister 'Mum told me that you were a bit, that you're not going to the doctor's although you're not feeling well. She said 'oh no, she always exaggerates, you know, it's a temporary thing', or 'I'm fine', you know, so I get the other side of the story.[23]

The hub network, while the most common form for those who use telephone and letters to keep in touch, is actually very fragile because of its reliance on one central person. When this hub is lost – most often as a result of illness, death or dementia – individuals in the family network have to reorganise their means of keeping in touch. In most cases, this will mean that migrants shift communications from their mother to a sister. So, for example, one Irish migrant explained that his telephone calls to his mother had become 'pointless' since the onset of his mother's dementia. Not only was he unable to learn anything reliable about his extended family network by questioning his mother, but also the extended family was no longer being informed of his news or even of the fact that he had called. Under these conditions, he was forced to locate a new hub for the network – a local sister who had taken on the role of primary carer of their mother.

Unlike telephone calls and letters, emails and faxes tend not to produce a hub style of network. Instead, they usually give rise to an 'all-channel' or 'distributed' network, 'in which communications proceed simultaneously in more or less all directions' (Urry, 2003, p. 160). This pattern is facilitated by the ability for identical emails and faxes to be sent with one click of a button to a large number of recipients. This style of network has the effect of increasing both the quantity of communications and the number of people involved in communicating. When families use email, in particular, we found that children of the migrants in Australia were more often encouraged to maintain their own contact with cousins and other kin in the home country. In the home country, much wider networks of kin (including nieces, nephews, aunts, uncles, siblings and cousins) are included directly in the regular exchanges of information, along with greater numbers of quasi-kin, friends and acquaintances.

This transformation of the communication network has important implications for who is involved in 'doing kinwork'. Contrary to di Leonardo's (1987) findings regarding telephone calls and letters, kinwork facilitated by email is much less likely to be 'women-centred' (Yanagisako, 1977). Otherwise silent members of the family – especially brothers, sons, grandchildren and cousins – use email to communicate across distance when they would not normally have participated in the routine exchange of telephone calls or letters. This may help to explain why, in our study, both men and women – fathers and mothers, daughters and sons – talk about their contributions to the work of maintaining the family. There are still some gendered dimensions to this work. For example, some parents are grateful that their daughters-in-law maintain contact, as their sons are inexpert or disinterested in this aspect of kinwork, whereas we did not encounter any examples of the reverse, when sons-in-law undertake this responsibility for their wives. However, the participation in kinwork does now seem to be distributed more evenly than was the case in the past. Indeed, several parents talked about how they were glad to see their adult children contact each other directly on a regular basis by email, thereby reinforcing and sustaining much stronger and broader kinship networks. As one parent said, 'things that we know, that we think they [the migrants] don't know, then they do know; they are informed'.[24] Such knowledge no doubt contributes to a sense that more kin can be turned to in times of need.

Emergent ideas of distance and proximity

In addition to changing the configurations of the networks of communication, it is apparent that different communications technologies provide different opportunities for perceiving or ignoring distance. For example, the simple fact of the elapsing of time that is required for letters to be exchanged has the effect of reminding letter writers of the distance that separates them. New, instantaneous, technologies of communication, on the other hand, give the impression of an immediacy of connection that belies the reality of geographic separation. Indeed, the role of time in the perception of proximity should not be underestimated (Urry, 2000). For example, the lack of a time zone difference between Singapore and Western Australia means that this group, in particular, is able to construct a strong sense of proximity and availability of kin, such that they are in contact 'all the time'. Time is also significant in other ways. An Irish woman talked with delight of how she was able to 'see' her newly born niece 'even before she was born', the result of the instantaneous nature of email technology in combination with the 8 hour time difference between Perth and Ireland. This is in striking contrast with migrants of the 1960s and 1970s – and the refugees even today – who might wait for months for the next letter and photograph from home.

At the same time, it is important to recognise that different frequencies of communication appear for all families at different points in the life cycle and in relation to different events. The weekly telephone calls and letters and daily emails are characteristic of general patterns of 'routine' contact. It is uncommon that a telephone call will be made between Europe and Australia on a daily basis, or that an email will be sent once a week on a particular day of the week. However, there are some occasions when the routine patterns of communication are disrupted in favour of a different – usually more intense – pattern of communications. In particular, 'crisis' moments in which the family is being reconfigured, such as in the case of births, deaths, marriages and especially the times of migration and visits, tend to evoke much greater frequencies of communication. Telephone calls, emails and faxes are often exchanged at a much greater rate in the lead up to and immediately after such events. These communications act to reinforce an even greater sense of the proximity of kin, during periods when the reality of distance might be experienced as particularly threatening.

The sense of time and distance also has a significant effect on the nature of the information that is exchanged during communications. Letters were often focused on telling the people back home about the strange places, people and practices in Australia. For example, an Italian woman wrote to her family about time she spent in Darwin, when 'we lived in the bush, with wild buffaloes, wild crocodiles and all sorts of things'.[25] A mother in Ireland laughed as she recalled amazing accounts sent by her daughter in Perth, who would:

> *write letters and tell us about these, I used to worry about these great spiders they would say were in the toilets, I think they had an outside toilet somewhere and they used to tell us about these deadly spiders that could be lurking there that would kill them, and about the mosquitoes and snakes ... and they'd send us letters about Christmas on the beach and all that, and it sounded very exotic.*[26]

Such letters often have the characteristics of adventure tales. Writing them down is no doubt important for the migrant, trying to make sense of their new surroundings. There is often an element of humour in the letters, possibly indicating a need for the migrant to turn what could be perceived as a crisis into amusing stories. The double distance – of the passing of time between the writing and the reading of the letter, and of the geographic space separating writer and reader – may serve to domesticate what are emotionally confronting experiences. By turning the crisis into an amusing story, the migrant may be better able to look on it calmly from the imagined position of their kin, at a safe distance in terms of both time and geography. Thus, letters appear to play an important role in curing homesickness for the new migrant. As an Irish woman explained, 'probably because I didn't know very many people [in Australia], I found writing letters [to people back home] very cathartic, really, and I would write a letter probably once a week'.[27]

In contrast, emails that are exchanged on a daily basis tend to be short notes passing on inconsequential information. The sense of the exotic that is associated with the passage of time and distance of the letter is replaced in emails by an emphasis on the immediate, the mundane and the 'everyday'. As a result, it becomes even more difficult to construct the distance that seems to be necessary for writing a letter; as one Irish migrant said, 'with email now, and the phone, [it is as if] not a lot really happens, because the minute something does happen, you tell them about it'.[28]

This sense of immediacy, arising from the availability of email, fax and telephone communications, has a significant impact on how the family network is experienced. In effect, it enables the erasing of perceptions of distance and a sense of continuity of existing family routines. For example, a Dutch migrant claimed that his parents are very knowledgeable about his everyday life:

> *They probably know more about me than they would have known if I lived in the area, because like you wouldn't talk about all those little details until you actually meet each other. And now I write to them [by fax] every week: on Monday we did this, and on Tuesday we did that, so they probably know more than they would have known ... we really give them a good understanding of how we live, and afterwards, in the faxes you can say 'oh we did this', like 'we bought a new buffet and put it in that corner'.*[29]

The exchange of the minutiae of everyday life is clearly important in constructing a sense of shared social space between physically distant kin. In line with past findings that suggest strong networks are important for the development of social capital (see Urry, 2002), this sense of knowing the details of family life from a distance is usually interpreted as a positive feature, through expressions such as 'we have always had good contact'.[30] As one mother in the Netherlands explained, 'I sometimes have this feeling that she has never been away, when I'm talking to her for half an hour on the phone; "catching up", she says. That is nice'.[31] For many, the ease of transnational communication means that distance becomes irrelevant to the family bonds of intimacy and involvement. As one mother in Italy claimed:

> *I can tell you for example where my son is right now, his whole life, whatever he does. I am not a possessive and obsessive mother, otherwise I wouldn't have enjoyed being in Australia [visiting him], but since we have this uninterrupted relationship of affection, we each look after the other.*[32]

As a result, migrants often feel like active, daily participants in the lives of family members on the other side of the globe. One Irish woman said:

> *When I came here first, it was that thing, you know, you'd get a letter and the news was sort of old, and then you'd respond to it, so it could have been a month. Whereas now [with email], I feel like I'm more involved in what's happening there. It sort of gives me a feeling of being part of it. Because sometimes you'd get news before other people, you know, who are there.*[33]

In times of crisis, particularly, this sense of connection can be especially important. A family in the Netherlands used email to keep their migrant kin updated and involved during the illness of a parent. They were able to 'keep each other informed by email every other day',[34] forging close bonds of support during an otherwise difficult time.

The overall effect of new communications technologies appears to be one of increasing the strength of the family bond. More people have become involved in communicating across distance, and communications occur more frequently. One effect of this growth in communications is the apparent erasing of distance as a factor in maintaining a sense of the family network. Licoppe (2004, pp. 135–136) argues:

> communication technologies, instead of being used (however unsuccessfully) to compensate for the absence of our close ones, are exploited to provide a continuous pattern of mediated interactions that combine into 'connected relationships', in which the boundaries between absence and presence eventually get blurred.

This sense of 'presence' in spite of physical absence is strong for almost all of the families in our study. Certainly, it is no longer easy to assume that physical proximity is necessary for sustaining intimacy or connection. However, it is also clear that access to communication technologies is uneven. As a result, some families find it much easier to sustain intimate relations across vast distances than do others. We now turn to some of the factors that affect the ability of families to maintain a sense of collectivity across geographic distance.

The capacity to communicate

There is often a sense of wonder about the incredible growth in access to communication technologies that has occurred over the last decade or so. However, our study clearly demonstrates that each form of communication technology excludes some people while incorporating others. For example, an important continuity between the letter of the past and the faxes and emails of today is the requirement of some level of literacy. For working class migrants and parents of the past, lack of access to education was a significant problem that inhibited their ability to write directly and fully to their distant kin. As a result, many letters were short, functional and highly formalised (Templeton, 2003). For the migrants in our study, literacy as such was not mentioned as a problem. However, all of our samples (except for the Irish and New Zealand groups) did report problems connected to the difficulties of maintaining more than one language.

In many cases, migrants have married monolingual English-speaking Australians, who are unable to communicate with their new parents-in-law. Some sons- and daughters-in-law make an effort to learn the home languages of their spouses, but this occurs infrequently and is usually insufficient to enable ongoing and in-depth communications with parents-in-law. Instead, communications tend to be mediated through the migrant. Non-English speaking parents talked of their concern about calling the households of their migrant children, where the English-speaking spouse or child might answer the call and create an awkward moment of miscommunication. Some migrants prepare translation cards and place them by the telephone so that English-speaking kin might let parents know when their child will be home to talk. But more often, parents simply stop calling rather than negotiate this possibility. The ramifications are usually not fully recognised until the birth of grandchildren. Suddenly, parents in the home country who are unable to speak English face the very real prospect of not being able to communicate with their own grandchildren. The heartache of hearing a grandchild's voice but not being able to understand may contribute to their choice not to call at all. Thus, the lack of a shared language, even when communication technologies are available, creates significant gaps in the family network and impacts on their ability to construct a sense of proximity.

But language is not the only factor that might contribute to such gaps. Other resources, too, might be lacking in particular family networks. Some people find that their time is limited by the responsibilities of work or childcare. For example, a Dutch woman said that, with two small children, her letter writing has become less frequent and enjoyable, so she has turned instead to the telephone as something that is less time-hungry:

> *I can't make time for [letter writing]. Then I really have to ask, 'do I feel like it now'. You really have to feel like it, otherwise you end up with a strange letter, where you'll think 'no, I can't send this!' Do you understand what I mean? That's just, no, it just has to be fun. And calling is often spontaneous.*[35]

For others, it is the telephone that is problematic. Fingers affected by arthritis find the buttons and dials of a telephone difficult to negotiate, and impaired hearing can be as limiting as the crackling or dropped lines of the past. At the same time, arthritis and impaired vision can limit the access that many have to technologies that rely on the written word, such as email, fax and letters.

In some cases, it is the power dynamics within the household that limit the effectiveness of particular communication tools. For example, in some households, one family member controls access to the telephone or computer. Fathers and husbands, in particular, were mentioned as expressing their disapproval of too-long or too-frequent telephone calls.[36] As we mentioned in Chapter 3, in refugee families, the husbands sometimes prohibit their wives from spending money on telephone calls on the grounds that the available limited resources should be directed towards supporting their own extended families, for which they have a culturally established responsibility. In response to this prohibition, some of the women secretly use housekeeping money to purchase a phone card. They then use a public telephone, rather than the home phone, to contact their families overseas. In other families, the problems are in the home country. For example, some migrants talked of not feeling able to speak to their fathers because their mothers would always answer the telephone and dominate the call. For others, it was not until the death of a father that they recognised his influence on communications with their mother. Suddenly, long, leisurely conversations would become possible on a much more regular basis.

In other cases, it is the perceived 'skill' of the migrant or parent in relation to a particular technology that creates a problem. For example, many people talked of just not feeling that they were good letter writers. For others, it was the style of 'chat' associated with telephone calls that was the problem. Thus, a Dutch migrant spoke for several people when she said:

> *I am not so good with the telephone, I have actually never had that so much. If I don't have a very good reason to call someone, if it is only to say 'hi' or 'how are you', then I am not very good at it.*[37]

For many parents, in particular, the telephone is still a tool reserved for passing on important information, and so a 'chat' is simply not comfortable. For example, a Singaporean mother said about her daughter: 'I don't normally call her unless I have something special to talk to her [about]. So she usually calls me'.[38]

With other communication technologies, it is lack of access to specialised knowledge and equipment that creates problems. People who do not or cannot keep up with newly introduced technologies are at risk of being left out of networks altogether, or have to find new ways of negotiating their access. Parents, for example, sometimes find that the introduction of new communications technologies results in the loss of their position as the hub of communication. Rather than being the direct receiver of letters and telephone calls, they depend on local children, and in some cases grandchildren, to provide them with printouts of emails, and to type and send their replies. In some cases, language-specific equipment is simply not available. For example, keyboards and computers that are able to write and interpret non-Western character fonts for Asian and Middle Eastern languages remain relatively inaccessible. With the growing use of mobile phone text messaging, the means of communicating and the power to communicate are shifting again. Already, it seems that those people who do not use mobile phones, or do not use the text function on their mobile phone, are at risk of being excluded from newly emerging networks of communication.

By far the most debilitating condition, rendering all communications unavailable, is the onset of dementia. As one migrant explained, she could no longer call her mother because 'the telephone has become an alien thing to her and she doesn't want it; she is scared of the telephone. So the only way I can be in touch with her is by writing a letter'.[39] Migrants facing the full effects of their parents' dementia talk of how, for them, their mother or father, while still physically alive, has already gone. It is impossible to construct a sense of proximity when the minds of the parents are not able to assist in that construction. While locally resident siblings can continue to interact with and care for their forgetful parents, the migrant child's relationship with their parent is fully conditional upon being able to imagine each other, and thus mutually construct the social relationship across distance.

It is clear to us that the most active family networks are those that have access to and incorporate the widest range of tools for communicating. In these families, the lack of access of some individuals to some technologies is compensated by access to other technologies. Also, the tendency for the network to rely on one, relatively fragile, hub of information is replaced by a vigorous exchange of information and expressions of relatedness across a vast number of nodes. In network analyses, it is understood that 'as the number of nodes increases, there is an exponential increase in the overall value of power of the network' (Urry, 2003, p. 162). For transnational families, this suggests that access to multiple means of communication serves to expand and strengthen the kinship network that each member might turn to for care and support. It is thus not surprising that so many of our interviewees commented on the 'miracle' of their experience over the past decade of an increased capacity to communicate. As one Dutch man explained:

> *The technical development is contributing enormously to simplifying that process [of keeping in touch]. Now you send an email or a fax, and in a couple of years you'll talk via a computer, then you can see each other and you can show things. It will make it all much easier.*[40]

However, it is also clear that not all families are able to take advantage of this technological revolution. A stark contrast to the miracle of communications expressed by the migrants in our study is the continuing struggle to communicate that was described by the refugees.

Global barriers to communication

For the refugee participants in our study, patterns of communication are relatively infrequent and irregular. This is not because of a lack of desire among the refugee family members to communicate with each other across distance. On the contrary, interviews are filled with the pain of separation and distance. Rather, this particular group faces two main difficulties. First, their geographic mobility is frequent and often sudden, as they respond to various dangers and opportunities in their countries of residence. This fractured migration experience makes it difficult to keep track of where kin are living at any particular moment. Typically, extended families are located across multiple nations, including Iran, Pakistan, Germany, Canada and Australia, often with several members unaccounted for. Yet, in spite of this dispersal and the uncertain lives of the family members, there is some level of continuous contact that ensures families retain a sense of collectivity. It is a testament to their commitment to the extended family that so many refugees sustain support and communications in spite of the difficulties they encounter. It is also, in part, a reflection of the relative absence of alternative forms of support from states and local communities. Second, and perhaps even more importantly, the countries in which the families of refugees live are typically characterised by poor access to reliable and cheap communication technologies. The services that most of the migrants in our study take for granted are rarely available to the refugee families.

It is notoriously difficult to quantify access to communication technologies, particularly when comparing statistics across nations (ITU, 2003). Nevertheless, the indicators that are available suggest that Iran has very low levels of access to technologies such as telephones and the Internet when compared to Australia and the other nations where the parents in our study live. For example, while in 2000 the other nations in our study had approximately 50 telephone lines per 100 inhabitants, Iran had only 16.9 (see Table 5.1). Similarly, an analysis of statistics by the International Telecommunication Union indicates that Australia and all of the migrant countries in our study score high in terms of digital access, with results ranging from 0.79 for the Netherlands to 0.69 for Ireland (see Table 5.2). Iran, on the other hand, achieves a much lower score, of 0.43 (ITU, 2003, p. 21). As this figure takes into account contributing factors

Table 5.1. Quantity of telephone lines by population

	Telephone lines per 100 inhabitants*
Australia	54.1
Iran	16.9
Ireland	48.5
Italy	47.1
The Netherlands	37.2
New Zealand	47.7
Singapore	47.1

*Data from http://cyberschoolbus.un.org/infonation3/menu/advanced.asp [accessed September 2005].

Table 5.2. Access to Information Technologies

	Digital Access Indicator*
Australia	0.74
Iran	0.43
Ireland	0.69
Italy	0.72
The Netherlands	0.79
New Zealand	0.72
Singapore	0.75

*Data from http://cyberschoolbus.un.org/infonation3/menu/advanced.asp [accessed July 2005].

Table 5.3. Call costs from a Telstra home line, July 2005

	Telephone call costs from Australia, per minute*
Iran	$1.35
Ireland	$0.21
Italy	$0.40
The Netherlands	$0.45
New Zealand	$0.21
Singapore	$0.45

*Data from http://www.telstra.com.au/phones/homeservices/distance_international.htm [accessed July 2005]. Call rates are the HomeLine Plus/ISDN Home/HomeLine Advanced 0011 per minute rates.

such as mobile and fixed telephone lines, adult literacy, availability of Internet service providers, number of Internet users and the price of Internet access, it represents a useful comparison of the level of access across the seven countries.

The inequalities of access are also clear in the costs of communicating from Australia with the six countries in our study. Although call rates do vary over time, there appears to be a consistent pattern in which Iran remains the most expensive destination. Table 5.3 gives a clear example of this difference in relation to Telstra services. This particular telecommunications company is not known for providing the cheapest service in Australia. However, as the previously national carrier, it does attract loyalty from many migrants (Baldassar, Baldock & Lange, 1999). It is also representative of the differences in costs that migrants and refugees encounter from most service providers. In July 2005 the cheapest per minute calls using Telstra were to New Zealand or Ireland, at only 21 cents per minute. Iran was significantly more expensive than all of the other countries, at $1.35 per minute. In comparison, sending a letter appears to be of equal and comparable costs for both refugees and migrants.

The relative equality of postage costs and access, especially when compared with Internet and telephone, helps to explain why most of the refugees in our study relied on communicating with transnational kin by writing letters. As one woman said, revealing a bag of about 10–15 letters, 'the only thing I do is write letters to them. I am going to post them all to the one address. Most of them live in the same house, and the others are living close by'.[41] When asked why the letters were not all placed in one envelope, as they were all being sent to the same house, the woman continued 'I have to write a letter for everybody individually. I cannot just write a letter to one and ask them to pass on my greetings to the others. They would feel insulted. They expect me to write a letter to each of them'. This example represents an interesting exception to the notion of the hub network of communication. Further research would be needed to begin to understand the specific cross-cultural and family dynamics that operate in examples such as this one. Some of the multitudes of letters are posted using the official postal service. However, not many people rely on this means of getting messages to their kin in Iran, Afghanistan or Iraq. As one man explained:

> *I used to write letters to my family, but not anymore. Once it took nine months before my letters reached my relatives in Afghanistan. Another time, a letter I sent with some American dollars to Iran was lost. Since then, I do not send letters.*[42]

There is a strong perception among the refugee families that the postal service in Iran is neither reliable nor honest, and they are disinclined to trust it. There is a sense that messages are intercepted and possibly even destroyed, that enclosed gifts are stolen and that in any case the delivery dates are much later than those advertised by Australia Post (see Table 5.4). During periods of civil war or other national emergencies, this sense of unreliability is further heightened. As one man said:

> *During the war people dispersed and many lost their lives, or went missing and could not be found by their relatives. It was not a matter of a phone call or writing a letter to each other. There was no telephone or postal service. You could only send a message through other friends.*[43]

Table 5.4. Costs and delivery times for standard letters sent through Australia Post, July 2005

	Postage from Australia, standard airmail letter up to 50 g*
Iran	$1.80, 6–7 working days
Ireland	$1.80, 4–5 working days
Italy	$1.80, 5–7 working days
The Netherlands	$1.80, 5–6 working days
New Zealand	$1.20, 3–4 working days
Singapore	$1.20, 3–4 working days

*Data from http://www1.auspost.com.au/pac/int_letter_select.asp [accessed July 2005].

The perceived lack of an efficient and reliable postal service means that most people rely on networks of friends and kin travelling in and out of the region to deliver letters, photographs and videos. One Afghan woman said that when her mother sends a letter to her, it takes more than two months to arrive: 'there is not a good postal service in my country'.[44] A man described how, if the official postal service was used, 'all the letters might be checked and even read before they arrive'.[45] But, in spite of the fact that they take long periods of time to arrive, and some may never arrive, it is clear that writing letters and receiving letters is just as important, if not more, to the refugees as it is to the migrants in our study. As one woman said, 'I love to see my mother's handwriting. I have read her letter many times, and I enjoy reading it every time'.[46]

The refugee group's practice of using travellers rather than postal workers reflects that of Italian and possibly other migrants groups in Australia up until the mid-1900s (Templeton, 2003). In spite of the apparent global spread of electronic communications, the methods of hundreds of years ago remain the most reliable in this context. Indeed, our interviewer was rapidly incorporated into this strategy of communication, her pockets and luggage stuffed full of letters, photographs, videos, gifts and some money to distribute to family networks in Iran. The request to carry such things is so taken for granted that, in one case when she had nothing from Australia for a family, she noted 'they wanted to know why their family in Australia did not write to them ... they were disappointed when they knew that the family did not even send a letter with me for them'.[47]

For the family members in Iran, any periods in which there is a complete lack of contact is very worrying – and the corresponding lack of financial support is potentially devastating. Given the economic conditions in Iran for the families of the refugees, it might be expected that our interviewer's suitcases were empty on the way back to Australia. However, this proved not to be the case. Family after family asked her to carry a small package or letters with her back to Australia, along with their good wishes and requests for support. She was also given many messages to pass on. From Australia, she took such messages as 'tell them that we are okay, but that life in Australia is not very easy – we want to help, but we have many expenses'.[48] From Iran, the messages included: 'please tell them in Australia that we need help, we can't pay for the children to go to school'.[49] The extended network of travelling friends and family is also important for finding out where people are living. For example, one woman in Perth asked the interviewee to see if she could get information about her husband's sister's son, who had dropped out of contact over a year ago.

While letters are the most commonly mentioned means of communicating for the refugee families, they are not the only strategy. None reported using emails or faxes, but many talked about telephone calls as important for maintaining contact. Nevertheless, such calls are strictly limited due to two intersecting factors: the high call costs, on the one hand, and the low incomes of the refugees, on the other. One man explained that when his father telephones him, 'he does not talk much as it costs him a lot to call, and at present he has not got enough money'.[50] Indeed, the most common motivation for making an expensive call from Iran to Australia is to request financial support – a difficult balancing act of the costs of the call against the potential aid it might return. Family members located in Perth, too, face high call costs that limit their ability to communicate with kin overseas. For them, the choice is often between spending money on a call to talk with their overseas kin, or sending much-needed financial assistance.

One way the refugees try to reduce their call costs is by purchasing phone cards. The cheaper rates on phone cards make the telephone more accessible to refugees, and help them to feel a stronger connection to kin overseas. For example, one Afghan woman said 'I call every fortnight. I get a phone card, some of them are cheaper than others. I always have one in my pocket'.[51] Phone cards have also become popular among the migrants, with Singaporeans and Italians, in particular, mentioning the good deals available from particular cards. Dutch fathers, on the other hand, take some pride in knowing which fixed line telephone company has the best call rates at any given time.

Implications of apparent proximity

The continuing inequality of access to communications technologies has a significant impact upon the capacity of some families to construct a sense of presence and proximity. The lack of a sense of presence, moreover, is highly likely to have a negative impact on the capacity of these families to exchange care and support, and to feel able to rely on each other for that care and support. However, presenting this lack of access as a problem assumes that all members of the family network have similar ideas about the level of support that should be exchanged across distance.

As we demonstrate throughout this book, not all members of a kin network share the same assumptions about who should be doing kinwork, or what constitutes a reasonable amount of kinwork to be done. Some family members chide each other with comments such as 'I was just calling to find out if you were still alive'. In several cases, both migrants and parents expressed their disappointment that they do not receive more attention from their distant kin. In one case, a mother lamented that her efforts to write letters and make regular telephone calls to her migrant son and daughter-in-law are not reciprocated. Her response was to stop communicating altogether; although this seemed fairer to her at one level, she is also angry and resentful that even this response has not incited her child to communicate more often. In another case, an ongoing dispute between two siblings means that a migrant daughter faces a significant crisis: her mother's mental health is declining, so that she no longer functions as a hub of communication, but her only sibling is still not prepared to communicate with her.

Even in the case of the refugee families, who have the lowest capacity to communicate, different expectations regarding kin support can lead to conflict. Just as some husbands try to limit the amount of money their wives spend on keeping in touch with family, it is clear that some wives try to limit the amount of money that is spent on the husband's family. For example, one woman talked about the conflict that regularly occurs when her husband's kin call their home:

> *I sometimes prefer not to talk to my husband's relatives when they call, they are so demanding and sometimes disturbing ... They think we are so rich in Australia, which makes me angry. The other day when my husband was talking to them on the phone I asked him to let me talk to them and explain our situation. He didn't let me, he didn't want me to upset his mother.*[52]

It is also the case that not all communications are equally satisfying. Indeed, many people talked of telephone calls that are stilted, letters that are upsetting and emails that stop communications altogether because they cause offence. While the breaking off of all contact is an extreme response and not especially common, the gradual diminishing of communications over time is a much more frequent phenomenon. In many cases, people become involved in local activities and think less often or with less pain about family overseas. The use of communications technologies feels increasingly inadequate for maintaining emotionally close and intimate relationships. As one Irish migrant said:

> *It comes in fits and starts. I might talk to them all a couple of times in one month, and then there'll be a dead patch. We're probably in a dead patch now, but before I spoke to my sister a lot. That kind of waned, but then with me going back there that got us back in touch, but it's actually quite hard because if you don't see them for a while then you haven't got any common thread any more, you know, you lose that gossip, like the little things, the banter and that.*[53]

Conclusion

It is becoming apparent that global mobility and the emergence of new communication technologies require people to develop new skills for maintaining and constructing social networks and relationships. Proximate residence and physical co-presence are no longer sufficient or necessary for participation in communities, if they ever were (Urry, 2000). Now, communal activities and identities can be constructed across geographic distance and under conditions of physical absence through the use of a variety of communication and travel technologies (Urry, 2003). Yet, at the same time, new skills also have to be developed to create a sense of distance and disconnectedness. The disassociation of physical presence from social presence presents new problems: it is now essential for people to make themselves socially 'present' and 'absent' in new ways.

Some scholars rightfully fear that individuals incapable of becoming 'present' in cyberspace will face new inequalities of access to citizenship (Urry, 2000). Indeed, this is not just an issue at the level of global and national politics. It is becoming apparent that physical proximate communities and intimate relations, too, are relying on the use of communication technologies to sustain a sense of 'presence' under conditions of physical absence (Wellman & Haythornthwaite, 2002). Indeed, the capacity to sustain a presence across geographic distance may be transforming the ways in which social relations are experienced and practiced, connected and disconnected.

In this chapter, we have demonstrated how transnational families have adopted new communications technologies as quickly as they have become affordable and available. In the process, these families have transformed the ways in which they think about the time and distance that separates them. In most cases, this has enabled the family to be imagined as a proximate, supportive network that transcends the realities of geographic distance. In some cases, this proximity presents new types of problems, which are worthy of future research attention. However, the greatest heartaches and inequalities are usually associated with an inability to communicate. The unequal access to communication technologies creates significant sources of conflict and anguish for refugee families. The inadequacy of existing communication technologies is a source of great pain for those migrants whose parents no longer have the capacity to sustain communications with the available technologies. Finally, the lack of a common language renders some relationships empty of content and satisfaction.

The problems faced by refugee families can only be addressed by macro-level changes in policy and distribution of resources, something we discuss further in Chapter 7. However, the fractures in the family network that result from factors such as dementia, illness and language barriers can be resolved to some extent by the families themselves. One important strategy is to supplement 'imagined' proximity with the real thing, by travelling to enjoy periods of physical co-presence. It is to this practice, of visits, that we now turn.

6
The Role of Visits

Because emigration in Ireland was a thing, you were gone and never seen again. But [now] not only [are] we, the younger emigrants, coming home [to visit], but now the parents are able to go and see them in their new environment ... so it probably satisfies them. They come out here and they see where you're living, and it's, 'Oh god, they're just living like us back in Ireland but it's in a different place'. They can see, and because if they had never seen, say the house where you live and the way you live, or what you do each day and all that, I suppose then they'd always be wondering. When we used to ring mum ... I'd say, 'I'm in the kitchen now', so she could picture us inside the kitchen on the phone, she could visualise the house. ... [After my mother moved to a new house] for years I hadn't seen my mother's home, where she lives ... it used to break my heart that I couldn't imagine where her bedroom's going to be or her lounge room, like was the kitchen okay for her? I found that very disconcerting. So when I went back last year and just walked in the door, to see it and feel it, oh god, it was great! I felt really at ease then, you know, at ease with myself. That was very disconcerting not being able to see.

(Irish migrant daughter)

We concluded the previous chapter by highlighting how transnational caregiving is reliant on the exchange of communication. 'Keeping in touch' primarily took the form of 'hearing' from loved ones 'voice-to-voice' over the telephone, or via the virtual 'voice' and 'presence' of letters, emails, faxes and text messages. Our study raises the question, however, whether long-distance communication alone is sufficient to sustain satisfactory experiences and practices of care.

As explained in Chapter 2, our methodology involved interviewing people in their countries of residence. We generally visited and talked to parents after having visited and interviewed their migrant children in Perth. In this way, we became an avenue for (corporeal) communication exchanges and co-presence, albeit by proxy, between transnational family members. While we ensured that all interview information remained confidential, particularly within families, we were frequently asked to carry assertions of love and care across national borders, in particular the declaration: 'tell them that we think of them, always'. The squeeze of the hand and the tearful gaze accompanying such statements gave a clear message: while staying in touch through communication technology is important in maintaining family networks, the ability to 'see' kin and to share a physical connection with them remains of great significance.

The expressed need, or at least the aspiration, for migrant and homeland kin to '*see*' each other 'face-to-face', to be physically co-present with each other, emerged as a significant feature of our study. Periods of co-presence and visits punctuate the years of separation and become 'key moments' in joint homeland and migrant histories, reinforcing, both with joyous reunions and sometimes painful tensions, the unity of transnational social fields. In this chapter we identify different types of visits and their role in transnational caregiving. We analyse the practice of visits by examining their frequency, timing and pattern across the life cycle. We interrogate the meaning and motivation behind visiting by attempting to understand the processes that inform the decision to visit. Finally, we consider the management of visits, their difficulties and challenges and the strategies people employ to try to ensure they have a 'good visit'. We also question how important the visit experience is to the practice of transnational caregiving.

The practice of visits: Form and function

To begin with, we can make some general observations. Visiting, in both directions – by migrant to homeland (migrant visits) and by parents to migrant (parental visits) as well as visits by other kin – took place in all sample groups. However, for some families, the visit remains more of a myth than a reality, being mainly longed for or intended rather than actually achieved. Even so, visits appear to be an integral part of the transnational care process, as they are of the migration process more generally (see Baldassar, 2001). While each visit is a composite of a multitude of meanings and motivations, it is possible to identify some specific types.

Table 6.1. Types of visits

Type of visit	Visitor	Role of Visit
Routine	Commonly migrants, occasionally parents or other kin	Staying in touch Maintaining family and consociate attachments Maintaining employment, profes sional or investment responsibilities; Often the visit forms part of work-related travel
Crisis	Parent or migrant or other kin	Provision of personal (hands on) and practical care, e.g., serious illness, divorce, sudden deaths, difficult births, change in level of dependence Parents often visit soon after migration and around the time of births Migrants often visit as parents become less independent
Duty and ritual	Parent or migrant or other kin	Rites of passage: (births, deaths, christenings, weddings, funerals, anniversaries) Fulfil family obligations
Special purpose	Parent or migrant or other kin	To 'see' for oneself how kin are Relieve homesickness, reconnect with place
Tourist visit	Migrant, parent and other kin, particularly grandchildren	Holiday with stopover to see family Maintain consociate and contemporary attachments Reconnect with 'roots'

As with all ideal types, the categories identified here are primarily useful as conceptual tools to further our understanding of the importance of visits in the process of transnational caregiving (see Table 6.1).

Types of visits

Routine visits are most common for people who can visit frequently and regularly, and often involve maintaining employment, professional or investment responsibilities. For example, one migrant returns to Italy routinely for his business. Another visits Holland to update her qualifications as a language teacher. Similarly, professionals often seek out conferences or business opportunities in locations near to their families living abroad (see Baldock, 2000). Migrants like these use their work-related visits as an opportunity to catch up with family. In this way, routine visits are characterised by the general activities involved in caring about distant kin and are not associated with any special or particular motivation to visit, other than the general desire to 'see' and 'be with' family. Some migrants (who can afford to) conduct routine visits purely for the purpose of staying in close contact with family. This is particularly common among migrants with children who have not yet begun school; one migrant explained: 'we go [to Italy] every Summer holiday so that the kids can play with and get to know their cousins'. Such routine visits are written into the contracts of the Dutch expatriates, who, as a consequence, are among the most regular visitors in our samples.

Crisis visits usually involve the specific need to care for distant kin, often through the provision of 'hands on' personal care or respite care, and usually in response to a sudden emergency (death, serious illness, difficulty surrounding birth of new baby or divorce). One woman, for example, returned to care for her mother who had developed a serious mental illness. She explained that the combination of relatively poor levels of psychiatric services, community/cultural stigma associated with mental illness and the denial of her siblings, kin and friends, made it imperative for her to become the 'responsible one' and 'go back and sort out this issue myself'. Her Australian-born husband followed her back after three months and found work for the duration of their stay. It took six months for a proper diagnosis of her mother's condition and a further six months for her health to improve to a level where the daughter felt able to return with her husband to Australia.

Duty and Ritual visits include attendance and participation in rites of passage (births, deaths, marriages), key celebrations and anniversaries, where the visitor feels an obligation to attend and where their physical presence is expected and anticipated. These visits can be differentiated from other types of visits by the fact that if the visitors lived locally, they would definitely have participated in the event, for example, attendance at funerals. Some of these visits might be viewed as a 'duty' and could conjure a sense of burden and reluctance to undertake the visit. This is the way King, Warnes and Williams (2000, p. 158) have described what they call 'duty visits' – as stressful and unenjoyable. In contrast, we found that *all* visits involve an element of stress and burden, something we discuss as an aspect of the management of visits later in this chapter. By duty and ritual visits, we mean the visits that people feel compelled to make; some visitors are very keen, others less so. The most common examples of such visits are those motivated by key life-cycle events including, in particular, weddings, funerals and special anniversaries.

Special (purpose) visits are undertaken for specific reasons: in particular, bereavement, the final stages of terminal illness, times of transition when elderly parents need to change their living arrangements and the birth of babies, especially the first-born. Special visits are also undertaken to ease the heartache of being separated from parents/children and grandchildren and to relieve migrant homesickness. Visiting for the purpose of curing homesickness is particularly overt in the Irish sample; as reported in earlier chapters, two-thirds of the migrant respondents had returned for lengthy periods. Migrants from other samples also articulate the value of a visit in this regard. Other reasons include to take a holiday after a serious illness or accident, to reconnect with place and national identity and to participate in special festivals including religious feast days, home town patron saint days and pilgrimages. Visits undertaken by people who do not visit often, perhaps only once or twice in a lifetime, also have the characteristics of special visits. Of course, these motivations could also form part of routine, ritual and other types of visits. Indeed, elements of each of these motivations might be present in any one visit. But 'special' visits are described by the visitors as being somehow different and set apart from these other motivations.

Tourist visits are brief visits to kin as part of a longer itinerary, which include travel to tourist sites. These visits are mainly conducted by extended kin, including adult grandchildren, aunts, uncles, cousins, nieces, nephews and friends of family. Included in this category are package tours that take groups of travellers to rediscover their origins (as in the case of Italian regional government tours for second-generation migrants who travel to Italy to visit their ancestral hometowns) or to discover the adopted homelands of their emigrant kin (for example, the tours organised by the various *italiani nel mondo* associations that take Italy-based travellers to visit the countries to which their kin emigrated).[1] Travel to attend class reunions, common among the Dutch, and to attend national or local festivals, like the annual Singapore Food Festival held each July, are other examples of this kind of visit. While our study did not focus on these visits, it is useful to make mention of them here. These visits invariably result in a broadening or widening of the transnational networks of caregiving, as migrants develop and consolidate relationships with a variety of kin (and vice versa), forming the basis of new or reinvigorated transnational communication relations.[2]

Timing of visits

The particular place of visits in the process of ageing is contextualised in this chapter within the broader histories of migration throughout the life cycle. While for some, visits may occur only months apart, others wait almost a lifetime. In general, visiting becomes more frequent and assumes a greater focus in the lives of both migrants and their parents at three key stages of life and migration: the period immediately following migration; time around the birth of babies and early childhood; and as parents become old and frail. These are the main 'crisis moments' identified in Chapter 4 as attracting the most frequent exchanges of care and communication. Visits take on a special quality when parents are welcomed to their migrant children's new homes, or when grandparents meet their grandchildren or when visitor and visited alike have to assume that they may not see each other again. In many ways, the role of the visit in transnational caregiving would seem to be most clearly defined when it involves these first and final meetings.

A key finding of our study was not so much the different types of visits that people engage in, but rather the considerable number and regularity of visits, both of which reflect the particular importance of staying 'in-touch' or of intermittent co-presence (Urry, 2000, 2003). In each of the migrant samples, visiting takes place, on an average, once every two to three years with many migrants receiving visits from family once every three to four years. Three groups, in particular, engage in the most frequent visits: Singaporeans, Dutch expatriates and the more recent professional or what we might call more 'individually-oriented' or 'cosmopolitan' migrants (see Chapter 3) across all samples.

Frequency of visits: Degrees of transnational mobility

Overall, the Singaporeans are the most frequent visitors, with (often multiple) visits to and from Perth commonly occurring annually, as one parent explained, '[my migrant children] visit Singapore so often that they don't go sight-seeing, just visiting relatives'.[3] There are several factors that contribute to this frequency. Family members live the shortest distance, a 4-hour flight, away. This, together with the fact that they are the only destinations in the study that share the same time zone, means that visitors can arrive on the same day they departed and do not have to deal with the physically demanding impact of long flights and jet lag. Further, the cost of travel is markedly (about 50 per cent) less than flights to Europe, making it affordable not only to undertake more frequent visits, but also very brief visits, which may not seem justifiable for the more costly travel involved in the other samples.[4] In addition, as noted in Chapter 3, there is a strong (though recent) history of education and business exchange between the two places, with thousands of Singaporean students studying at Perth universities each year. About two-thirds of the migrant sample were originally international students who later acquired permanent residency, and many of their parents had also studied in Perth. As a consequence, there is a large though diverse 'Singaporean community' in Perth and migrants and their parents invariably enjoy wide networks of family, friends and acquaintances in both places, thus increasing the impetus to visit.

As important, if not more so, as these historical and structural factors are the Singaporean interviewees' shared notions of what they described as a cultural impetus to visit. Migrants and parents alike spoke of a heavy emphasis in Singaporean society on close family relations. Many referred to 'Asian values' and 'Chinese culture', in particular, as fostering an expectation that family members 'see' each other as often as they can.[5] A similar set of expectations defined by interviewees as cultural is also evident in the Italian sample, where an individual's sense of well-being is often associated with a sense of family closeness,[6] and among the refugees, where kin networks have become the primary social structure for individuals in the absence of well-functioning communities or states (see Loizos, 2000, p. 126).

Singaporean migrants were also the most likely of all the sample groups to plan 'tourist' visits around homeland festivals, possibly because Perth is close enough to justify and accommodate the time and expense.[7] Even so, Singapore–Perth visits are not without their impediments. Visitors from Singapore have to factor in the additional cost of health insurance as they are not covered by reciprocal health care arrangements. This is not an issue for the New Zealanders, who enjoy health cover and do not require a visa to enter Australia. They also live relatively close, with comparatively affordable flights (similar to Singapore) and a time zone difference of 4 hours, which is generally considered manageable even for brief visits.

Like Singaporeans, many New Zealanders seem to take for granted the possibility of seeing each other. However, New Zealanders in our study, on average, visit less frequently than participants from other groups mentioned above. The particular composition of the sample may contribute to this finding – interviewees were mainly young men and their parents are all relatively young and healthy (we discuss the impact of gender, age and stage in the life cycle on the practice, motivation and management of visits in more detail later.) Further, migrants and their parents do not generally perceive the move to Perth as a permanent migration, but more like an extended stay. This has an impact on the granting of licence to leave.

The two countries are perceived to be very close, geographically and culturally. For example, they share the ANZAC public holiday (established to commemorate the loss of Australian and New Zealand soldiers in World War I) and several trans-Tasman sporting events. Families feel they are in easy reach of each other and it is possible that the sense that they are close enough to travel any time results in New Zealanders travelling less often than those more distant migrants who need to carefully plan and save for trips. This said, many parents bemoan the fact that 'Perth is not Australia' and that other cities like Sydney and Brisbane are much more convenient to visit. They complain about the relative lack of direct flights and the need to spend long periods of time waiting for connections.

The finding that Singaporeans make the most visits would appear to reinforce the argument that *distance* is a key factor influencing the practice of aged care across borders. However, the comparatively infrequent visits by the (relatively proximate) New Zealanders, as well as the high frequency of visits of certain individuals in the European countries included in our study, belie this notion.

Visiting, and more generally, transnational mobility, is a relatively familiar aspect of the lives of most participants, particularly the professional migrants and most especially the Dutch expatriates, who expect to travel constantly as part of their career paths. As noted in Chapter 3, there is a general sense of taken-for-grantedness about transnational mobility and the possibility of making return visits that makes it easier for these migrants and their parents to cope with the distance. Some companies financed trips home as frequently as every three months. The professional, more 'individually-oriented' or 'cosmopolitan' migrants from all samples have lived all over the world and are seasoned tourists:

> *It was not a big shock for them, all my family move around the world. I am not the only one ... My cousins are moving around as well, [as is] my sister.*
>
> (Italian migrant son)[8]

> *They have a nice ideology, my parents: Three, four months at one place, move on to the next place. Because like I'm in Australia and my sister's in Hong Kong, my brother's in Singapore. So ... they think of well, three months in Australia, four months in Hong Kong, another five, six months in Singapore. And do that sort of rotational deal until someone gives birth to a kid or something like that. Then they would stay longer.*
>
> (Singapore migrant son)[9]

Something of the magnitude of this mobility, and its recency, is reflected in the following quote:

> *The fact is, I go to Italy every year and if there is ever an emergency or problem [I return immediately] ... the fact that I can go often I think reassures them. Because there is always this attitude that when someone goes overseas, they forget their family, their country ... There are people who didn't come back for 20 years, or didn't ever come back at all ... That's what I'm saying that there are two different types of migration; there are people who were forced to migrate, and people who migrated because they wanted to expand their knowledge, they wanted new adventures, new experiences. You have to distinguish between that.*
>
> (Italian migrant son)[10]

Clearly, the migrations of professionals or 'cosmopolitans' are not associated with the classical humanitarian and labour migration images and experiences of loss and rupture. Rather, they provide vivid images of transnational social fields that are well travelled and frequently criss-crossed:

> *The first one that came over was my sister and her husband and two children. We arrived in May and they came over in December. The next year my parents came and my mother-in-law, and it kept on going that way. It's rare that we don't have visitors in a particular year.*
>
> (Dutch migrant daughter)[11]

> *It's quite an organisational effort too ... My [in-laws] live in Melbourne and ... in Sydney so, ... we try and go through Sydney and stay with them for three or four days and you go to New Zealand and then come back to Melbourne, stay with them for a week and come back to Perth. It's sort of like a 'round the family-tree trip.*
>
> (New Zealand migrant son)[12]

In large part, the extent of this transnational mobility has been brought about by improvements in the technology of travel and its increasing availability and affordability. Particular travel technologies characterise certain eras of migration, making historical context another important consideration in understanding the frequency and duration of visits. When people had to rely on the costly and time-consuming journey by boat, visits were infrequent and longer, often six months or more. As travel has become more accessible and time efficient, visits have become more frequent and people are prepared to travel even for short visits of a week or so. Further, with the rise in travel for tourism, increasing 'tourist' visits (encompassing both sightseeing and time with family) are being made by migrants and their parents, as well as by extended kin and friends in both places, in particular second-generation 'non-migrants' (that is, the nieces and nephews of migrants living in the home countries) travelling for a holiday or on honeymoon.

The development and expansion of both travel (from boat to plane) and communication (from letter to email) technologies have arguably also increased the desire and motivation to visit. If migrants and their geographically distant kin are in more constant contact, they are more likely to engage in transnational caregiving, which appears to influence the pattern of visits.

Pattern of visits: Between family life cycle and tourism

We found that for the longer-distance professional migrants (Dutch, Italians and Irish), stage in family life cycle combined with stage in the migration process tend to define their pattern of visiting. Generally speaking, during the early stages in both the migration and family life cycles, this group is more likely to receive visits from homeland kin. Visits from parents, particularly mothers, also increase around the birth of grandchildren, events that usually occur soon after migration, when grandparents are still quite young and mobile. Parents' greatest impetus to visit is to nurture relationships with grandchildren and to 'be there' for children when needed. It is often during this early stage in the migration process that migrants themselves cannot afford the time and money to visit, or are subject to visa restrictions, for example, awaiting permanent residency status. As children and grandparents grow older, migrants with young families tend to take it in turns with parents to visit regularly to ensure that grandparents and grandchildren develop close relationships. This middle stage of the process is characterised by the migrants' children becoming more autonomous and beginning to visit on their own, and migrants beginning to undertake visits without them. The late stage in this process occurs when migrants' parents become too old to travel and migrants increase the number of visits to participate in their care. We attempt to capture something of the pattern of visits in Table 6.2.

This broadly drawn pattern of visits across the lifespan, punctuated by crisis, duty, ritual, tourist and the odd special visit, can be summarised as beginning with more parent visits, followed by a period of turn-taking, where routine visits are exchanged between parents and migrants, and ending with more migrant visits. Of course there are

Table 6.2. Pattern of visits

Time of visit	Visitor	Type and role of visit
Usually early in migration history	Parents and sometimes other family members	**Special visits:** • First parent visit • Reassurance that migrant is okay • Acceptance of migrant's migration • Reconciliation
Usually early- to middle years of migration history	Migrant	**Special visits:** • First return visits • Decides where to settle • Relieves homesickness • Atonement for leaving **Routine visits:** • Maintains family/consociate attachments • Maintains work/formal connections
Early- to middle years of migration history and/or early in migrant's family life cycle	Mainly mothers and sometimes fathers and other family members	**Duty/Ritual visit:** • Birth/care of grandchildren **Special visit:** • Establish relationship with grandchildren **Routine visits:** • Support migrant family • Consolidate family relationships
Middle years of migration history and/or of migrant's family life cycle	Migrant	**Special visits:** • Visits with spouse/children • Relieves homesickness • Atonement for leaving **Routine visits:** • Maintains family/consociate attachments • Maintains work/formal connections **Ritual visits:** • Rites of passage (e.g., birth, wedding)
Middle years of migration history and/or of migrant's family life cycle	Parents and/or other family members	**Duty/Ritual visits:** • Rites of passage (e.g., wedding, anniversary, graduation)

(*continued*)

Table 6.2. (*continued*)

Time of visit	Visitor	Type and role of visit
		Special visit: • Maintain family relations and duties and fulfil obligations
Middle years of migration history/life cycle	Various family members	**Routine visit:** • Sustain family relations **Tourist visit:** • Holiday **Special visit:** • Attempted migration **Duty/Ritual visits:** • Rites of passage
Middle years of migration history	Migrant	**Duty/Ritual visits:** • Rites of passage
Later years of migration history and late in parent's life cycle	Migrant Second and subsequent migrant generation	**Special/Crisis visit:** • Care for ageing parents **Special/Tourist visits:** • Rediscover family roots • Develop consociate relationships

Notes: Tables assume migrants migrate when young, consistent with the majority of our samples.

numerous variations on this general model. For instance, the annual visits home for the Dutch expatriates dominate their visit pattern. We found that if migrants can afford to visit more, parents may visit less, and vice versa. In general, parental visits to migrant children tend to be less common than visits in the other direction. This is especially true for refugees, New Zealanders and the post-war Italians. Some parents fund the visits of their migrant children and grandchildren rather than make a visit to Australia themselves. The usual argument is that it is more important for the migrants to visit the homeland so that the entire family network can reunite, rather than just the parents and migrants spending time together. However, if there are few relatives in the home country, or if their only grandchildren live in Perth, parents are likely to undertake more visits than their migrant children. The most frequent parent visitors in all samples are mothers, especially widows who appear to have more autonomy and time to travel, although their fares are often paid for (in whole or part) by their migrant children.

The importance of visits: Meaning and motivation

By far the most striking feature of caring for and about distant kin through visits, across all sample groups and relevant to all types of visits, is both migrants' and homeland-based parents' strongly expressed need to 'see' their loved ones, 'with my own eyes', in order to confirm their health and wellbeing 'for themselves'. The apparent universality of this need leads us to conclude that a primary role of the visit is for the visitor to 'see' and so validate the status of the visited; to check they are 'really okay'.

Seeing is believing

'Seeing with your own eyes' ensures people can reliably and comprehensively assess the wellbeing of a distant family, something that most interviewees felt cannot be effectively done through the various forms of long-distance communication technologies. These concerns reflect the common custom of hiding the truth from family members in order to protect them. As discussed in previous chapters, people are often trying to 'read between the lines' of letters or they are intent on 'listening to the sound of the voice', as well as to what is being said on the phone, to try to discern if the true state of affairs is being hidden from them. Asking other locally based contacts to 'play detective' and check up on their distant kin, 'in the flesh', is a common strategy employed to restore confidence. However, none of these seems as effective in reassuring family as an actual visit.

As noted above, our data show that the need to undertake a visit in order 'to see', to check the wellbeing of distant kin, is most strongly expressed by parents soon after their child's migration. This is perhaps because most parents have little, if any, access to information from other people about their migrant children. Only the Afghan and Iraqi refugees, some Singaporeans and the few post-war Italian and Irish migrants in the sample belong to 'communities' in Perth that include wider transnational networks extending to the parent's hometown. Other visits that feature this motivation 'to check up on', over and above the usual desire (that characterises all visits) 'to see' loved ones, are visits by grandparents to 'see' grandchildren, particularly new babies, and by migrants to 'see' ailing parents, particularly if disability (dementia, deafness, paralysis) makes it impossible for them to communicate by phone or letter. Importantly, a significant outcome of 'seeing' children, grandchildren and parents, is that visitors feel more reassured and more accepting of the migrant's decision to settle abroad. A visit appears to be the best way of resolving tensions associated with a lack of license to leave as the quote at the top of this chapter and those following reveal:

> *I made this trip and saw my daughter's life with my own eyes. I feel much better now. (Refugee mother commenting on her visit to Iran to see her daughter)*[13]

> *They found it really hard to accept. It took them a long time so they obviously didn't want me to go because I would have been so far away from them. For the first year, my mum would cry every time someone mentioned my name! Just the feeling emotional, 'oh, that daughter, I wonder what she's doing now. I don't know, I can't see her'. It's not being able to see, not being able to verify with her eyes. When they came over, they saw how my life was and they could verify things, it reassured them.*
>
> (Italian migrant daughter)[14]

> *When my sister went back after visiting, she was probably talking about it all the time, and so my mother said 'I'm going out to see for myself!' [laughter] I think she just more or less felt, to satisfy her mind, you know what I mean, we were bringing up our family here, and I think she wanted to see for herself, you know, the kind of life that we'd chosen. Yeah ... [later in the interview reflecting on the impact of the mother's visit] Anyway, she was ah, very happy.*
>
> (Irish migrant son)[15]

> *They were here for four weeks, with us, in our house. Especially for them, because they, they had never flown, it was really an experience, right. And they loved it, everything they saw here. Because my mother, when we left my mother said, 'oh, now you're going to that country and it's a cowboy country' and she had the strangest ideas. And then it is good they came over, because they can see it, they see what your life is like, they see your environment.*
>
> (Dutch migrant daughter)[16]

'Seeing' each other also reassures people that their kin 'haven't changed a bit', that they remain 'just the same', that despite the distance, they still have a son or daughter, mother or father. In this way, attachments with significant kin are revived, and in the case of grandchildren, formed and developed. These attachments ensure people feel they are maintaining a sense of 'closeness' with significant kin, that they still 'know' each other. People also spoke about 'getting to know' kin and developing relationships that carry obligations for a continued set of interactions. As mentioned in Chapter 5, visits often inspire an increase in transnational communication both before and immediately following, and visits also often increase and broaden the transnational network of kin and friends. The maintenance of these attachments is often underpinned by a sense of moral obligation towards family, evident in duty and ritual visits, through attendance at funerals, baptisms, weddings, significant anniversaries, Christmas, class reunions, times of crisis (migration, birth, illness, death) and the sense of duty of seeing family regularly.

Being there

While it might be assumed that the visit is an opportunity for the visitor to 'care for' the visited through the provision of 'hands on' care, it is also, in and of itself, an expression of 'caring about'. During the transnational visit, any distinction between 'caring for' and 'caring about' collapses, because just 'being there' is often viewed as an enormous expression of care, even if the visitor does not provide any of the specific types of care (personal or practical) normally associated with proximity. Further, it is in the act of 'seeing' (in the flesh, as opposed to in photographs) and of 'being co-present' that we find the specific relevance of visits to the emotional and moral support of transnational caring, even though these types of caregiving are the very ones that can be provided over distance, without being physically present. As Hage (2003) has noted, sometimes the greatest gift we can give someone we love is the gift of our presence, of simply 'being there', as when we say with heartfelt sincerity, 'thank you for coming' or 'thank you for being here'.

The need to 'see' each other and to be 'co-present', expressed by our informants as a key feature of visits, supports Urry's (2003, p. 164) claims that 'co-present interaction is fundamental to social life', a point we return to in Chapter 8. Urry (2003, p. 163) highlights the importance of travel in enabling co-presence and producing intermittent moments of physical proximity to particular peoples, places or events:

> This mutual presencing enables each to read what the other is really thinking, to observe their body language, to hear 'first hand' what they have to say, to sense directly their overall response, to undertake at least some emotional work ... These social obligations are associated with obligations to spend moments of 'quality time' often within very specific locations often involving lengthy travel away from

normal patterns of work and family life. There is often a quite distinct temporal feel to the moment, it is 'out of time', separate from and at odds with the 'normal' processes of work, leisure and family life ('take time out' as the advertising goes).

The emotional and moral support that we defined in Chapter 4 as the bedrock of transnational family relations is rendered particularly special and out-of-the-ordinary when it can occur face-to-face rather than from a distance. The simple act of seeing each other and of being co-present sets the visit apart from other forms of transnational communication. The very fact that visits across borders usually represent time taken out of everyday routines can render them into powerful expressions of caregiving. In fact, for some families, visits provide a greater propensity for intimacy and the development of close family relations than is available to proximate family, as the following migrants explain:

> *I did find ... that they all enjoyed visiting here, and ... this made my relationship with each individual member of the family very different, and much – yes, how will I say this – much more intimate than they have with each other. Because they see each other, for example, on birthdays, on festive days, that sort of thing, but they actually never sit around a table together for a week in a row, where you start talking about quite different things, because you have much more chance to finally sit down with those people, and that is what they have with us, and never with each other [laughs].*
>
> (Dutch migrant daughter)[17]

> *I mean he doesn't want to worry me with his presence. But I said, no, you're my guest. I'll bring you around. I'll hire a car and bring you around and all that. To tell you the truth, I have never been like that with him in my life except for that.*
>
> (Singaporean, talking about his migrant brother)[18]

The idea that distance motivates people to appreciate their families more deeply was commonly expressed by those we interviewed. Similarly, the belief that 'distance makes the heart grow fonder' is evident in the way people talked about missing each other and longing to be together again. It is important to point out that the notion that seeing each other less frequently somehow leads to 'closer' or better quality exchanges, as implied in the above quotations, must be tempered with the possibility that people are on their best behaviour during visits. While our findings support Goffman's (1963, p. 22) view that 'copresence renders persons uniquely accessible, available and subject to one another' and although Urry (2003, p. 164) notes that 'eye contact enables and stabilises intimacy and trust, as well as the perception of insincerity and fear', we caution against the tendency to privilege co-presence as more authentic than other forms of communication by assuming that being present is directly related to knowing 'the real' or 'the truth'. This denies the possibility for co-present people to deceive each other, to suppress parts of themselves and to act in a role (especially for short periods of co-presence). An Irish migrant woman makes this very clear when she describes how she hid her unhappiness in Australia from her family in Ireland:

> *When I went home [to Ireland] for that six weeks I should have, you know, been able to express the misery that I was feeling, but for that six weeks I didn't. I kept the whole charade of everything being okay, I still don't really know why. I think I just so badly wanted it to work. ... I think, in retrospect, they must have suspected, but nobody at any point when I was there said 'is there something you want to tell us', you know, there was none of that.*[19]

Sense of place

Often each visit fulfils a multiplicity of needs and motivations. Some outcomes, like 'seeing' and 'being there' are manifest and actively sought, other functions of the visit are more latent and less overt. In addition to the conscious consolidation of consociate attachments of kin and family, a more generalised set of connections develop from attachments to place and the accumulation of knowledge about broader contemporary events.[20] Our interviews included questions about national identity and our findings indicate that visits have an impact on identity formation at many levels, including local, ethnic, national, global and 'diasporic'.[21] Migrant visits often include trips to popular tourist sites to enable visitors to (re-)connect with places of significance (at some/all of these levels). Another important aspect of visits is the attendant shopping sprees to purchase appropriate diacritical markers (tourist icons and fashion) which are the broadly identifiable symbols of ethnic and national identity back home in Australia. Similarly, when parents visit, they purchase mementos of their time in Australia to display back in their own homes. These signs and mementos prove the visitor has attained the status of 'one who has visited', and contribute to the development of ethnic, national, diasporic and/or 'cosmopolitan' identities. In this way, the visit provides migrants with an avenue for maintaining an ethnic identity in the Australian multicultural political climate (Baldassar, 2001). Similarly, the non-migrant often develops an identification with Australia. This creation of multiple or hybrid identities affirms ideas about ethnic and national identities being more than just a 'commodity within corporate multiculturalism' or 'currency within global human rights discourses', but also a way of 'providing some kind of ontological security in a fluctuating world' (Fortier, 2003; see also Giddens, 1991).

These consociate and contemporary knowledges and attachments are catalysts for transformation and authentification. People talked about 're-discovering their roots' and finding out about their past. In this way, the visit brings about a type of rite-of-passage where visitors begin to identify more with the places and nations they visited. The transformation in identity and newly acquired status is evident in, for example, improved language ability, greater awareness about the home/host country and an increased sense of belonging. The visit is also sometimes spoken about as affording a kind of redemption and spiritual renewal. Here the visit can be described as a secular pilgrimage – a personal quest for spiritual, emotional and mental wellbeing (Baldassar, 1998, 2001). This quest is expressed by informants as a set of physical, corporeal needs, a longing to revisit the places of their past, the places they knew so well but had been separated from. This need is only fulfilled by being physically present in a *place* where they can breathe or smell the air, touch the soil and taste the produce, hear the bell tower or the call to prayer, feel the earth and see the homeland. The visit is thus an antidote to 'homesickness' as it soothes the ache of longing and nostalgia.[22] Where kin connections have been dormant, these visits are an example of what Grillo (2007) calls an opportunity for 're-transnationalising', where consociate relations are reinstated and reinvigorated. In cases where there are no longer significant consociate obligations and attachments, as after the death of parents, migrants may continue to visit through secular pilgrimages, to renew and recharge their connections to place and their memories of loved ones.

The knowledge and attachments gained and developed through visits are important in developing the foundation for transnational caregiving relations. But what makes some people uproot themselves from their families to spend an extended amount of time providing personal 'hands on' care to new babies or to dying parents, while others do not visit at all? How do parents decide they need to visit their migrant children to check how well they have settled in their adopted homeland? How do migrants decide whether they should try to reach a parent's deathbed 'in time' or to attend their funeral? We now examine the complex mix of factors that inform, influence, facilitate and impede visiting across borders.

Deciding to visit: Negotiated commitments, capacity and obligation

Like the decision to care outlined in previous chapters, the decision to visit is influenced by three key variables: negotiated commitments (family histories), obligation (duty) and capacity (ability). Visits, as a feature of transnational caregiving, must be contextualised within family histories. Antagonism between family members, for example, might lead some people to choose not to have anything to do with 'distant' kin who they believe have not behaved appropriately. Although there were very few cases reported in our study, some migrants and some parents *choose* not to visit each other at all, or decide never to visit again. There were also a number of cases of people who are not able to make visits, which will be discussed later in the chapter. Suffice it to say here that while we cannot assume that not visiting or few visits indicate fraught family relationships, our findings do suggest that visits are an important (and possibly an essential) part of conducting care across borders and maintaining transnational families.

Capacity (ability) to visit

Visiting takes time and money, leaves the visitor's family of residence devoid of their usual contribution of responsibilities and services and removes the visitors from their support networks. However, the *capacity* to engage in distant care is not solely a function of available resources. For example, how much time and money a family have at their disposal to allocate to distant care might be quite different to how much they are willing to commit to transnational caregiving.[23] One Dutch migrant commented that her parents 'spent a lot of money on [visits]; because they are not rich but for them that was very important so they did spend their money on that'.[24] Those few Afghan and Iraqi refugees who manage to visit their families at all do so at enormous financial strain, having to spend all of their savings. In contrast, one Irish migrant explained that she had not visited Ireland because of the cost, preferring to prioritise her children's education and family holidays elsewhere.

The capacity to undertake visits is also influenced by the willingness of spouses and other family members to allocate the requisite resources; one Italian woman commented, 'after all, if you're not spending your money on a visit, you would be spending it on the family here'.[25] When the family budget is already deficient, visits may mean significant sacrifices for those who stay behind. As noted in earlier chapters, many Afghan and Iraqi women express dismay at their husband's reluctance to spend money on the long-distance care of their families. There were also instances in the other samples of individuals not supporting their migrant spouse's desire to visit, even though they could afford it financially.

Refugees have the most restricted capacity to engage in visits; they have the least finances, the most tenuous jobs and the strictest visa regulations. In addition, depending on the social and political stability of the localities in question, they may be placing themselves at risk by returning to their homelands or transit countries. This does not necessarily mean, however, that they do not engage in visits, or even the fewest visits. They also express some of the greatest need and obligation to visit, as is evident from the following case:

> *It was very scary. I knew I would be placing myself and my children in great danger but I could not wait any more. I went to visit my family and my husband's family in Pakistan first. Then we crossed the borders and went to Kabul by car. We arrived in my brother-in-law's house. They did not know about our visit beforehand. We had no phone and no contact prior to the journey. They were shocked when they saw me at the door. Then they sent my children to get my mother who lived nearby. The children told her who they were, that we had arrived from Australia. She didn't believe them. She came immediately and as soon as she saw me she cried and I cried and the children were crying. I stayed with mother for a few weeks, we gave thanks at a nearby shrine. She was worried whenever the children went outside to play. She thought that somebody might recognise that they came from Australia. I wore a full Jadori cover so that nobody could recognise me. When we left, I took some old family photos with me. We were searched at the border and my mother's friend who was travelling with us hid those photos and pretended to be an old woman.*[26]

One Iraqi refugee desperately wanted to visit his mother and his wife and children but had to wait over two years to get his Australian passport. He was warned by friends that he should not leave the country until he qualified for, and had received, Australian citizenship. His friends were afraid that permanent residency would not be enough to ensure his re-entrance into Australia. Eventually he could bear to wait no longer and decided to risk returning to Iran without an Australian passport. Against the advice of his friends he spent six months with his mother, wife and children in Iran not wanting to return to Australia until they had all been granted entry visas. Another Iraqi refugee couple was less fortunate:

> *Both of us. We both are so depressed and want to visit our family in Iran. We have never lived separate or so far from them before. Here, in Australia, we have material goods but our heart is not happy any more. Although we had many problems we were happier when we lived with our family. We could laugh from our heart. I could visit my family easily and used to go to the shrine everyday. We were always out in the market or in my relative's house. Here I am at home most of my time. I have nobody to talk to ... I used to call them every two weeks but I phone less these days ... not only because of the cost but I also get very upset whenever I talk to them. They talk about their problems and I cannot do much about them.*[27]

Further, many refugees experience 'survivor guilt' for having been lucky enough to escape the hardships of their homeland, only to leave behind dependent siblings and ailing parents (see also Ward, Bochner & Furnham, 2001). As a result of their relative good fortune, humanitarian migrants often feel an enormous sense of obligation to assist those who they feel they deserted.

The parents and families of refugees who remain in the original homelands of Afghanistan and Iraq or transit countries like Iran and Pakistan are usually unable to visit their kin in Australia and generally have great difficulty in obtaining the necessary visas. All the refugees in the sample aspired to be able to sponsor kin to Australia. Different people privileged different kin in this aspiration. For example, one woman wanted to sponsor her parents, but was restricted by her husband's family's preference to sponsor other relatives. For most refugees, sponsorship is, however, prohibitively expensive. There are also character, language and health checks that parents have to fulfil. As noted in previous chapters, refugees on Temporary Protection Visas (TPVs) are unable to visit family in home or transit countries, even in the case of serious illness or trauma because leaving Australia effectively terminates the safety provision of the visa. TPV holders are also ineligible for family reunion applications and while their family members back home qualify to apply for refugee status through the Global Humanitarian scheme, this is a process that can take many, many years.

Some Afghan and Iraqi refugees describe living any distance apart from their families and societies as having considerable negative social and emotional consequences. Some women, in particular recent Hazara arrivals, explained that they rely on their extended female kin, not only for childcare and domestic support, but also to accompany them in public places. Not having access to female kin drastically reduces the autonomy and agency of these women. In addition, marriage partners are often sought and arranged through family connections, so that the truncated family structures that result from forced migrations often have a negative impact on a migrant's marriage prospects. In this context, undertaking visits to be with parents and extended kin can be emotionally and socially liberating for both visitor and visited. The refugees we interviewed are clearly attempting to fulfil their traditional roles and duties across borders. It was not uncommon, for example, for single men and women to travel back to the transit countries of Iran and Pakistan to look for a suitable spouse among relatives and within the displaced community:

> *I decided to marry very recently. I wished to marry an Afghani woman. My sister and I went to Iran to see whether I could find some good girl suitable for me to marry. My parents knew my wife's parents very well. After we made some enquiries, we found my present wife's relatives in the city of Tehran. We went to visit them and there I met my wife for the first time. I liked her and her family. Then I expressed my interest and she also agreed to marry me later on.*[28]

In many cases, the original homeland is not accessible or there are no longer any family members living there, or those things associated with 'home' no longer exist, including entire villages. Many refugees had been forced to sell property in order to raise enough capital to leave the country (including costs for human smugglers). As a result, 'home' becomes associated with family members wherever they are located. In this context, 'being with' family members is likely to be as close as refugees come to feeling 'at home', and families spend years saving the necessary funds to undertake visits to 'be with' kin. In addition, families (particularly of the more recent arrivals) are dispersed throughout the world such that their networks of caregiving and support extend beyond the classic binary home–host country contexts. We found that the refugees in the study remain in regular contact with kin in a number of transit and host countries and that visiting these places is just as important to them as visiting 'home'. Many refugees choose to travel back to a transit country (Pakistan and Iran) rather than the original homeland; it is considered safer than trying to go home, they usually have more relatives there than in the homeland, and they have often spent several years living there, so that it is considered an 'adopted' home (see Kamalkhani, 2004).

This said, the original home country remains significant both emotionally and politically and most refugees never give up hope of being able to safely return there one day, although this desire often becomes a 'myth' of return, especially as children are born and the costs of visits become even greater. For example, one family who had lived in Iran for five years before settling in Australia in the mid- to late 1990s took eight years to raise the money needed to send the mother and a daughter to visit Iran and Herat. They found none of their relatives were left in Herat, most were living in Iran and others were dispersed in Germany, Holland, England and Canada. The following year they travelled to Canada to visit another daughter and sister. Another couple had hoped to travel to the USA together to visit the wife's ailing mother. Due to financial constraints, the wife ended up having to go alone. As she anticipated, her mother died a few days after her arrival. This woman expressed great relief and happiness at having been able to participate in her mother's funeral. Ideally, refugees would like to be able to take their whole family with them to visit their kin. Deciding which family members can afford to go is a source of great anguish, as the following quotation makes clear:

> *Oh, how can I leave my children here? They have nobody to take care of them here. They also wish to see their aunty. I don't enjoy going alone. How can I enjoy travelling on my own and leaving my children here on their own? I want to save enough money to be able to take all my children with me. ... Each visit to my sister's costs us lots of money, actually all our saving for many years. ... I can travel without my husband but not without my children. My sister wants to see them all. They have children their age and they want to see each other.*
>
> (Iraqi daughter)[29]

As the refugee case makes clear, visas and migration policy restrictions as well as issues surrounding the political security of countries have the most severe impact on frequency of visits as, unlike other constraints (time, money, employment and risk or perceived danger) they are generally not negotiable. Migrants face significantly fewer structural impediments. The New Zealanders, as already mentioned, are the freest in this regard as they do not need a visa, while visitors from Singapore have to accommodate the additional cost of health insurance. Apart from the refugees on TPVs, migrants awaiting permanent residency are also unable to leave the country and there are also restrictions placed on migrants with spouse visas. One Italian migrant who received a carer's pension to look after her invalid husband was not able to leave Australia for more than three months as this would terminate her pension. These and other policies affecting transnational caregiving, including community and family initiatives, will be further examined in Chapter 7.

While capacity to visit is clearly most restricted for refugees and their families, visits by migrants and their parents are affected by many of the same factors. In summary, capacity to visit can be analysed as constraints around *movement, time, money, employment and sense of security.* Finding the time to visit is often as difficult as finding the money. Employment flexibility is primarily an issue for migrants in full-time work, and proves to be a bigger impediment for men than women, an issue we return to below. Setting aside some time to visit is usually possible. However, getting enough time off work to make the visit 'worth the money' is a problem that many migrants and parents confront. A majority of the longer-distance migrants expressed the view that unless they have several weeks at their disposal, the visit is not really economically viable. Time allocated to visiting is also often time away from paid employment and time away from the family left behind.

It is important to note that, increasingly, perceptions about security, danger and risk, particularly in the political climate of post-September 11, impact on people's sense of capacity to visit. Many people expressed concern, beyond the fear of flying that incapacitate a number of interviewees, that travelling is becoming more dangerous. The security risk associated with visiting was most clear in the case of refugees attempting to travel to war-torn locations, or crossing borders that might ultimately block their return. Yet, even migrants in our sample expressed fears about their personal safety travelling in a world defined by 'war on terror'. As the following quotation by a Dutch migrant makes clear, it is often a sense of obligation that ultimately informs decisions about spending the time and money, and taking the risk, to visit:

> *After the London bombings my friend decided, that was it, she wasn't going to travel out of Perth anymore. I said, "well, that's fine for you to say, but I have family living abroad and I don't have that choice".*[30]

Obligation (duty and felt need) to visit

While being able to afford the 'costs' of travel is an important factor determining *capacity* to visit, it is profoundly influenced and mediated by the sense of *obligation* to visit. The *obligation* or sense of duty or felt need to visit stems from a range of motivations influenced primarily by cultural attitudes and expectations around caregiving as well as available services in the home country. For some (both refugees and migrants), the sense of moral responsibility and guilt for being so far away, is, in part, appeased by making visits home.

The visit becomes a particularly important form of care in times of crisis. Our findings suggest that there are two particular stages in the life cycle where the physical presence of transnational kin is often anticipated and even expected across all samples. Many migrants hope that their parents, especially mothers, can be present around the time of the birth of a new child to provide extra support and care. Most parents want their migrant children with them when they become old and frail or critically ill. There are different cultural constructions of both independence in old age and of new parenthood, as well as diverse cultural notions about preferred care arrangements at these times, which have important implications for the visit trajectories of both migrants and parents (see Baldassar, Wilding & Baldock, 2006). Nevertheless, when parents become frail or disabled, and when a new baby is born, the type of distant care that kin can provide is affected. There is often an increased need for personal, 'hands on' care which triggers a greater frequency and duration of visits, particularly by women, because this kind of care can only be provided during visits. In the case of severely disabled or critically ill parents, the objective is often to provide respite care to siblings and other carers, as much as it is to have contact with parents.

Compared to the Dutch and New Zealand cases, the relative lack of available and affordable aged-care services in Italy and their total absence in Iran contributes to the sense of obligation to visit felt by migrants and refugees from these places. A sense of obligation is also evident for the refugees, Italians, Singaporeans and Irish in the cultural expectation that children provide 'hands on' personal care for their elderly, manifest even when services are available. This sense of obligation is further enhanced by the strong sense of family shame associated with the use of institutional care reported by the refugees, Italians and Singaporeans. In these cultures, the use of institutional care is popularly characterised as 'locking' older people away. As a result, Italian migrant women are more likely than the Dutch, for example, to undertake visits to provide personal care to their parents.

In the Dutch case, the availability and preference for support services provided outside the family results in personal care not being required of Dutch migrants, although local siblings may provide a great deal of such care. While the Netherlands has the best and most extensive aged-care services of all the countries in the study, this does not mean that Dutch migrants visit any less often than others in the study. Rather, they simply are not as involved in hands on personal care (Baldassar, Wilding & Baldock, 2006). They express their moral and emotional support in other ways, including by just 'being there'.

Some migrants decide to return to the home country because, either there are no siblings, or it is clear that none of their siblings are able to provide the care necessary to meet their parent's care preferences. A number of factors impact on this sense of duty, including the gender of the migrant, whether they have any particularly relevant caregiving skills, and their work and family commitments. These factors inform the negotiation of commitments within families. For example, despite having several capable local siblings available to provide care, an Irish migrant nurse undertook a lengthy home visit to look after her mother who was dying from a terminal illness. Her professional skills and her availability made her the most appropriate person to do this. For other migrants, such a decision is neither necessary nor expected due to the extensive support networks already in place. Of course, the presence of such networks can create its own problems for the migrant in terms of access to parents, as will be discussed below.

The migrant's sense of obligation to care for ageing parents is mirrored somewhat by parents' sense of duty to assist their children with the care of new babies and growing children generally. Italian, Dutch, Singaporean, Afghan and Iraqi parents, in particular, reported that their identity as 'good parents' depends on this. Many parents feel compelled to visit after the birth of grandchildren, particularly if the mother is thought not to be coping well. Italian and Singaporean parents tended to assume that their presence would be needed, anticipated (or at least hoped for) and accepted. All the refugees expressed a strong wish that their parents could be with them. One Dutch (grand)mother organised a visit at a moment's notice when she received a call from her daughter asking for help with her newborn. The Irish appeared to be the only group where there was no explicit, taken-for-granted expectation that parents visit for the expressed purpose of assisting with newborn grandchildren, although there were individual examples of this view in other groups, for example, a Dutch mother who categorically stated that she visited to spend time with her daughter and not her grandchildren. We also have some evidence to suggest that Irish migrant women, unlike women in other samples, prefer to travel back to their homeland after the birth of their babies in order to be 'looked after' by their mothers. Further research is needed to confirm and explain such differences in visit practices as these findings may be an anomaly of our particular sample.

Women across all samples are more likely to be involved in visiting, especially for personal (hands on) care purposes. Women are more likely to be in a position to find the time needed for these visits; they are usually earning less than their spouses and many choose to take part-time and casual work specifically to enable visits at short notice. This does not mean, however, that women's employment does not suffer as a result of taking time off to undertake visits home. We encountered several cases of women who made career choices to accommodate their visits. Some lost their jobs and others had to forgo certain entitlements, like long-service leave.

That women's employment is likely to be more flexible as well as considered more expendable than men's are undoubtedly relevant factors. But equally so is the fact that women are more often involved in caregiving across all cultures (and this is what places them in the more flexible, expendable jobs in the first place). They are also more likely to feel confident that they have the necessary caregiving skills. Arguably, there is greater cultural expectation for women to visit in order to assume this role than for men. Even in the refugee, Singaporean and Italian cases, where men do express a strong sense of cultural duty to care for their ageing parents, it is often their wives who travel to provide the actual hands on personal care.

The obligation to provide care through visits is usually met with love, willingness and sacrifice. This said, transnational caregiving is certainly not devoid of difficulties and tensions, many of which find expression in the management of visits, to which we now turn.

The management of visits: Governance and control

So far in this chapter, we have tended to focus on the positive aspects of visits. We have argued that the visit, in particular the ability to see each other and to be co-present, revitalises, strengthens and perhaps even sustains, transnational family relationships. Visits often provide special shared times that deepen and strengthen family ties. Indeed, our findings indicate that for most people, most of the time, visits are a quintessentially good experience and people greatly enjoy them. However, another important dimension of the visit experience is its potential for disappointment and disillusionment. We have abundant evidence to show that visits can be problematic and are not necessarily the solution to tenuous or difficult family relations. Perhaps part of the negative potential of visits can be attributed to the high expectations that tend to surround them; as with ideas about family, they are steeped in expectations of good will and happiness. The time, effort and money it takes to make visits can leave people anticipating wonderful moments of joyous reunion and resolution. The reality may be quite different, particularly when it involves crisis visits and caregiving, although problems are common in routine visits as well. In this section we deal with the vexed issues of accommodation, autonomy, access and reception during visiting.

No visitor is immune to the heartache that visiting can bring. As mentioned in Chapter 4, one migrant came face-to-face with a dementing father who actually forcefully removed her from his home. Members of a refugee family were distraught to find their entire village destroyed. A few migrants intend never to visit again because they find the pain of departure too great. Some parents feel they have no space to themselves when their visiting migrant children use their house 'like a hotel'. Several migrants wondered whether it would be better if their parents stopped visiting, because they found their negative attitude towards living in Perth very upsetting. And visitors often feel hurt and neglected when their hosts, busy with their own lives, are too preoccupied to give them their full attention.

Often a great deal of pressure is placed on the visit, because it is precious time. Visitor and visited alike want it to be perfect, as evidenced in the following quote:

> *If you come all this way [even if it's a work trip] you have to [find time to spend with your parents]. Then mum [was upset]; 'why did [you] come if you are so busy here and so stressed?' 'I'm sorry I'm so stressed but I'm stressed because you are here and I can't see you.' I had to work from 8–8. There was this deadline. ... I was probably quite snappy in that week because I was just torn between work and having a few days to spend with mum ... So we probably didn't plan that very well.*
>
> (Dutch migrant daughter)[31]

Many migrants actually bemoan the fact that they never have a 'proper holiday' because they feel constrained always to use their leave time to visit home. Some parents too commented on the way their visits become working holidays as they assist with practical tasks and babysitting.

The stresses of the visit can be very draining and, for the migrant, include learning to fit back into a less familiar lifestyle, coping with the changes in health and wellbeing of loved ones, being identified as a migrant/foreigner rather than a local, and experiencing the loss of autonomy in the face of the demands of the host. For those migrants who have spouses or children who cannot speak the local language, these factors are exacerbated. Mostly migrants struggle with the desire to spend as much time with their parents as they can even though much of this time might be boring, bothersome and banal. Parents experience the stresses of migrant visits too, through disruptions to their routines.

Parental visits to Australia are also not without difficulties. Lifestyle, language and stark differences in climate and diet contribute to a sense of alienation for those parents who cannot speak English and come from cultures quite different to Australia's. These parents often resent the resulting dependency, isolation and sense of bewilderment and hopelessness. Lack of English language renders parents especially dependent on children for all their activities, placing an enormous burden on children's time:

> *Look, when mother visited here, [my husband] was unemployed, I was busy studying, so we had a limited income, secondly she came for two months, and that is a very long time, especially under those circumstances, and thirdly she speaks not a word of English, while I have absolutely no acquaintances [who speak her language], not for years ... So I actually did never speak [the language] and furthermore, I did not know anybody, so all the people who visited here, or wherever we went, I always had to be with her ... Actually, you got sick of one another.*[32]

The family of refugees usually have limited opportunities to visit and for many, the only way they can see family abroad is to migrate. For many elderly, this is not always a positive decision:

> *For old people it is better to stay in Iran than migrate to Australia. Here is not good for them. They get depressed and worry about dying in a strange country. This is what happened to my father. He died in the USA soon after he left Afghanistan. He was a writer and poet. He had a big library at home. My brother took him to the USA. He felt very sad there. He told him once that he had brought him to a 'golden prison', 'why did you do that to me?'. My father said that he did not want to leave Afghanistan. He died a few years later away from home.*[33]

If parents are already sceptical about their child's migration, and have not easily accorded them license to leave, the visit may make them prone to highlighting everything that is 'bad' about Australia; 'the sun is too hot, the people are too cold, the food is too bland, the shops are too boring, how can you bear to stay here?'.[34] These stresses and strains sometimes contribute to migrants and parents renting apartments to ensure autonomy or travelling together with other migrants who provide a kind of psychic buffer (Delaney, 1990; Mandel, 1990).

It is important to reiterate that migrations often result in fractured family histories. Even the most conscientious transnational communicators cannot stay fully cognisant of every aspect of the lives of their distant loved ones. There is often considerable tension and competition (both direct and symbolic) between non-migrants and migrants. For their part, the stay behinds (who may actually be repatriated after a migration of their own) can harbour a variety of perspectives on the migrants. At one extreme, they may view the migrant as a kind of 'lucky fortune seeker' who benefited from the opportunity to leave the homeland resulting in the non-migrant being left behind to care for property and kin – a view that can foster jealousy and resentment. At the other extreme is the image of the migrant as an 'unfortunate sacrifice' who was forced to leave, whether due to national or family poverty or simply through poor personal decision-making – a view that often fosters guilt (Baldassar, 2001). For their part, the migrants may begrudge their stay behind kin the additional support they receive from parents who are proximate and therefore more able to provide childcare and other kinds of practical, hands on care. Alternatively, they may feel indebted to them for taking on the lion's share of the burden of caring for matters back home. Finch (1989, see also Finch & Mason, 1993) points out that distance is sometimes seen as a legitimate excuse which absolves translocal family members from the expectation to provide care. As we discussed in Chapter 4, we did not find many examples of this. Nonetheless, limiting visits can be a way of using distance to avoid transnational caregiving duties. These ideas and feelings about distant kin, the 'stuff' of negotiated commitments, whether true or imagined, colour and shape the quality of transnational relations.

Duration of visit: Like fish in the fridge

One of the biggest issues for visitors and visited alike concerns the duration of a visit: what is the right amount of time? The following sentiment was commonly expressed by migrants and parents alike:

> *Yes, we do have [a house]. ... The first couple of years we stayed over. Without children at first, then with one child, and then the second was on*

its way. And then we realised that that was not a good idea. That is ... you can stay with someone for a few days, but not a couple of weeks. Fish will be fresh for three days.

(Dutch migrant daughter)[35]

The ideal length of time for a visit appears to vary between country groups. For example, Dutch migrants commonly express the view that visits longer than six weeks have the propensity to become difficult and stressful, while Italians more generally feel that the longer the visit, the better. This does not mean that Italians reported less difficulties and stresses associated with visiting; rather, they tended to conceptualise 'Italian culture' as valuing time with family no matter how difficult. For example, several women described 'being Italian' as involving expectations that you are always available for your family and that family comes first before anyone else, including yourself.

In general, migrant visits are shorter than parental visits, lasting on average one month, except in those special visits undertaken for the explicit purposes of providing care. Parental visits tend to be longer, lasting at least one month and often more. New Zealanders and Singaporeans are the most likely to undertake visits of under three weeks.

Particularly important to the subjective experience of the visit and its management are the personal family and migration histories of the people involved and their impact on overall transnational family relationships. In Chapter 1 we noted the importance of the migrant's 'license to leave' in these migration histories. Our data appear to indicate that in those cases where migrants were not easily afforded license to leave, transnational relationships, including visit experiences, are more fraught. In a related fashion, if migrants feel their migration has not been 'successful', they are less likely to make visits. Some people feel that visits can have a negative impact on settlement and that if migrants or parents visit too often, the former's ability to settle in the new country might be inhibited.

Autonomy and accommodation

The very thing that makes the visit special is also what makes it difficult. Unlike local kin, who can visit each other and return to their own homes, transnational visitors are generally accommodated in someone else's home. Many respondents commented on how welcome they are made to feel as house guests. However, in a few cases the partners of migrants expressed some resentment at the 'mini invasion' they felt the visitors represented. Similarly, negotiating space for new partners is not always easy. Several migrants returned to their childhood home to find their parent's new partner having 'the run of the place'; this made them feel like 'it wasn't really home anymore'. Others are anxious about taking their partner on a visit, especially if they have no, or limited, ability in the home language. Parents, while they love to have their grandchildren visit, are often unaccustomed and easily overwhelmed by the noise and mess created and the labour-intensive routines required.

The tensions associated with sharing accommodation are often exacerbated by the need to provide care to ageing parents. Local carers, usually the migrant's siblings, may view the arrival of the migrant as an opportunity to take some hard earned respite, leaving the visitor to take over the caregiving needs for the duration of their stay. This is especially common when ailing ageing parents are accommodated in an extended family setting, for example, where the parent has moved in to live permanently with a child or where a child has inherited the parental home along with the primary responsibility of caring for their parents as they age. This particular care arrangement not only has an impact on the autonomy of the parents, who have to surrender governance of the household to the child they are living with, but also a significant impact on the migrant child, who no longer has direct access to their parents' home (Baldassar, 2001, p. 30). For example, in order to be able to communicate and visit with her terminally ill mother, one Italian migrant daughter had to rebuild her relationship with a previously alienated sister who had taken over the care of their mother in her own home.

The sibling in whose home the parents are living theoretically has complete control over access between the parents and the migrant child. A similar degree of control is available to those proximate sibling(s) who have taken on the primary responsibility of care for parents still living in their own homes. The level of authority held by siblings can significantly impede the ability of the migrant child to care for their parents from a distance as well as the extent of their influence on care-related decisions. As contact between migrant children and parents increasingly has to occur through an intermediary, relations between geographically distant siblings often become more intense than ever before.

Similar issues about governance and control are pertinent to situations where the ailing parent is cared for by their spouse as the following case makes clear:

> *It has more to do with the fact that she [the mother] was the boss and that everything had to happen through her. It went this far, that last year when I went home [to visit] my father, in hospital ... my mother fixed everything,*

> *'You should sit here now, and dad over there, in that wheel chair ... now we have to go home ... and now we have to leave'; so much so that I thought: 'Tomorrow I want to go on my own. I want to talk to my father, I want to talk to my father without my mother' ... And then I went there myself, the next day. [My mother] didn't know ... And then I was there all day to talk to him, and it was so wonderful even now I'm so intensely happy that I did that.*[36]

Most migrants and refugees are likely to be accommodated with family on visits and, as noted in Chapter 4, parents across all samples are generally accommodated in their children's homes in Perth. Extended visits, particularly of large groups, sometimes involve rental accommodation. There is some evidence to suggest that hosts take offence if visitors on very short visits (in either direction) do not stay with family; similarly, for longer visits there is an expectation that the visitors will stay with family for at least part of the time. For example, an Irish mother was upset when her son stayed with his sister rather than with her for some of the visit, although the son thought he was relieving the burden on his mother by giving her some time alone in the middle of his visit.

The management of visits is often a difficult and exhausting task. However, visit experiences also change over time, with patterns developing that often increase the pleasure of the visit for all concerned, as the following Italian migrant daughter explains:

> *My visits get better each time. Because I have less expectations of – I just expect things more naturally and as they come. I remember the first time – Oh you know, I'm not going to see my mum [enough]. I have to do this with her, I want to do this with her, we have to – you know, get close, we have to – and this time, I just didn't care. Whatever we did was fine as long as we spent time together but – and maybe they're getting better because there's less, I don't feel those pressures that I felt before, about me going back to Italy.*[37]

Some families become more accustomed to their transnational responsibilities over time and get better at them. The question remains, however, how necessary are visits to successful transnational caregiving?

Conclusion: The necessity of visits?

Visits are important in establishing, maintaining and reinvigorating family relations, in reconciliation and acceptance of change, in fulfilling family obligations and duties and in the myriad issues implicit in the relationship between identity and place. That migrants tend to visit more than their parents appears to indicate that ultimately the obligation is on the one who leaves to maintain their ties and connections to people and place, particularly as parents age and can no longer facilitate transnational relations. A key question to arise out of our analysis of visits in the process of aged care across borders is: Does a family's sense of 'successful' transnational relations require mobility/visits?

As discussed in the previous chapter, there is a view that the revolution in communication technology has contributed to collapsing the sense of distance, making people feel they are closer. The impact that advances in technologies have had on the regularity and frequency of visits is an interesting question. The common assumption, as intimated by the following migrant, is that such technologies might replace the need to be co-present:

> *Yes, the contact with the parents, when you compare it with the fifties when a lot of Dutch people came here, and they had to write a letter to the Netherlands and it took six weeks. And when you compare it with nowadays, where you can just send an email through the internet and over a couple of years, because we have a little [digital] camera now ... But that's a matter of a couple of years and then everybody's doing it. So the technical development is contributing enormously to simplifying that process [of migration]. Now you send an email or a fax, and in a couple of years you'll talk via a computer, then you can see each other and you can show things, it will make it all much easier. It will make the world a lot smaller.*[38]

Urry's (2003) notion of degree of 'meetingness' is useful in analysing the significance of visits and the role that co-presence plays in transnational family relationships. Urry hypothesises that the basis for 'physical, corporeal' travel are new ways in which social life is 'networked' (he refers to the 'networked family'). His argument tends to be ahistorical in that it implies social life was not 'networked' across distance in the past, but he links this 'new' mobility inextricably to the information revolution. He suggests a fundamental difference in the function of visits in recent times compared to earlier periods and that visits or moments of co-presence 'are essential for developing those relations of trust that persist during often lengthy periods of distance and even solitude' (Urry, 2003, p. 163).

Our findings suggest that new technologies increase the incidence of visits. By increasing the sense of 'presence' of family across time and space, communication technologies also appear to increase the desire for intermittent co-presence (see also Urry, 2002, 2003). At the same time, advances in travel technologies have increased the opportunity – and thus the pressure of obligation – to visit. In other words, the more communications occur across distance, the more likely people are to undertake or at least desire to undertake to-and-fro travel. It does not seem useful to argue about which is more important; the distinction appears arbitrary as the families in our study make use of all the forms of 'staying in touch' across distance that are available to them.

Indeed, what we observe in our study is the emergence of a new set of factors being taken into account in the process of negotiating commitments to care across transnational borders. In addition to past histories of kin relations, it appears that relations to the material world – of technologies and resources – are just as important in determining perceptions of the obligation to care. That is, they impact upon capacity to care, which in turn informs the sense of obligation to care. Access to the material world is highly influenced by national and international institutions, structures and organisations. It is to these issues that we now turn, in our discussion of policy and its implications for transnational caregiving.

7
Implications for Policy

I took up Australian citizenship as soon as I could. Until 1992, I would have had to relinquish my New Zealand citizenship. At the end of 1992, the Act changed and I took out joint citizenship immediately. I had it by the middle of the following year. I would not renounce my New Zealand citizenship. But I've always, whenever I'm traveling overseas I only ever travel on a New Zealand passport because until very recently Australia was not a party to all the conventions. As a New Zealander I didn't need a visa for virtually anywhere in the world. Albania was one of the few countries to need a visa, and who'd want to go there? Whereas, to enter the United States until very recently, or to enter Britain or France, Australians need a visa. So I use an Australian passport to enter and re-enter Australia, but then travel elsewhere on my New Zealand passport.

(New Zealand migrant son)

When you have no passport, you have no power.

(Afghan refugee son)

Across the world today, millions of people provide informal care to parents and other family members. In Australia alone, 13 per cent of the population, or 2.6 million people, are defined as carers, that is a 'person of any age who provides any informal assistance, in terms of help or supervision, to persons with disabilities or long-term conditions, or older persons (aged 60 years and over) ... ongoing, or likely to be ongoing, for at least six months' (ABS, 2004, p. 71). A narrower definition, which focuses on the notion of a 'primary carer' who 'provides the most [ongoing] informal assistance, in terms of help or supervision, to a person with one or more disabilities' (ABS, 2004, p. 77), still generates a count of approximately 474,600 carers, about one in every five carers (ABS, 2004, p. 50). The majority of these primary carers live with the main recipient of their care. Such local carers are rightfully given close attention in today's policymaking. However, they are not the only ones involved in caregiving. There are other, translocal and transnational carers, who either provide help and support from a distance within the same country, or give care across national borders. These translocal and transnational carers have remained largely invisible in policymaking.

In Australia, carers can receive financial support in the form of a *Carers Payment*, with 71,691 recipients in 2002, and a *Carers Allowance* – 307,694 recipients in 2002 (Fine, 2004, p. 222). Also, paid carers' leave can be negotiated as part of Workplace Agreements; a test case in 1994/1995 established a minimum entitlement of 40 hours of paid leave.[1] As we will show, translocal carers within national borders receive some support from community organizations. Would transnational carers benefit from any assistance of this nature?

Transnational migrants participate in caregiving across national borders in four main ways. Firstly, they seek to provide emotional, moral, financial and some practical support through transnational communication; this requires the ability to communicate readily and cheaply across borders with ageing parents and/or their local caregivers. Secondly, they make return visits to give emotional, practical and personal care to their ageing parents in times of need. This means they must have the opportunity to travel. Thirdly, they may seek to bring their parents to Australia in order to care for them on a long-term basis. Finally, and usually as a last resort, some transnational migrants may repatriate to their homeland to care for their parents there. Each of these strategies is subject to specific obstacles, some imposed at the macro level of government and industry, others occurring at the meso level of the community. In this chapter, we ask whether it is possible to overcome any such obstacles through changes in policy and practice. Alternatively, are policymakers, employers and community service providers able to assist transnational caregivers through the introduction of new policies and practices that target their specific problems of transnational caregiving?

We discuss such questions with reference to the four avenues that transnational migrants have for participating in transnational caregiving. Much of the information we provide relates specifically to Australia and so will be of particular value to migrants and refugees residing in Australia and their transnational family members. Nevertheless, it is important to emphasise that answers to the questions we raise are relevant *anywhere* across the globe. None of the issues is unique to migrants and refugees who have settled in Australia. Any transnational caregivers with parents and other relatives across national borders encounter them and seek their resolution within a framework of capacity, cultural obligations and negotiated family commitments.

At the same time, the extent to which the concerns of transnational caregivers can be resolved depends on macro-factors, which can only be dealt with through changes in social policy at an institutional level. It is here that we may encounter differences in opportunities for transnational caregiving between nations. With the exception of New Zealanders, the migrants and refugees we included in our study came from countries that do not have economic, political and legal frameworks in common with Australia. There are some bilateral agreements, for example, regarding health insurance and the possibility of dual citizenship, but international visa regulations are strict, and there are a number of other macro-factors that inhibit opportunities for transnational caregiving. Throughout the chapter, we draw comparisons in this regard with the situation in the European Union. Transnational migration, including migration by retired people, is on the increase in Europe (King, Warnes & Williams, 2000). Such migrants move between countries that share economic, political and legal frameworks. In this sense, migration between EU countries may be deemed *translocal*, rather than transnational. Although the nations that form part of the EU do not as yet have the same welfare provisions and opportunity structures, there is a common resolve at governmental level that any inequities for people moving between EU countries should be minimised (see also Ackers & Dwyer, 2002). Such assumptions cannot be made for transnational migrants and refugees who moved to Australia, except in the case of New Zealanders, and it is these issues we discuss in this chapter.

Transnational communication and contributions to care

Informal carers who have primary responsibility for their elderly parents are well aware that there are a number of major components to caregiving that require close attention. For example, even before any hands-on caregiving takes place, *local caregivers* need to gather information about the problems faced by their care recipients. This may entail gaining data on medications required, and about appropriate doctors and specialists. Caregivers may need to accompany their elderly relatives to medical appointments in order to get such information, or consult medical dictionaries and websites. They will then also need to find out what care arrangements are required and available in their community, and decide on what needs to be done in terms of housing. The actual implementation of such care plans can be a time-consuming and complex matter, especially when it involves a change of residence (such as when an elderly parent moves in with children or to a nursing home) and/or specific arrangements for domiciliary nursing care. Often, there are also financial and legal arrangements to be made. Local carers need to assess the costs of domiciliary or institutional aged care, and frequently negotiate with paid caregivers to ensure the most cost-effective care. If they are to manage the financial affairs of the person they care for, they also need power of attorney.

We found in our study that many transnational migrants try to participate actively in these aspects of caregiving from a distance, through letters, emails, faxes and phone calls to parents, siblings and neighbours, and in a few instances even through direct communication with medical doctors and other paid professionals. In so doing they gather information, help decide on care arrangements and accommodation and provide emotional and moral support; the latter may also involve their children staying in touch with their grandparents. Further, migrants are able from a distance to provide financial support for their parents' care, as in the case of Singaporeans and refugees, communicate to their relatives abroad that they wish to forgo their inheritance to offset the costs their local siblings incur in caring for parents (we found some Italian examples of this),[2] or help manage their parents' finances, as was done by some New Zealand migrants. However, transnational migrants do not usually seek power of attorney; this is generally given to local siblings or other relatives who do the everyday caregiving. Migrants who are only children may visit to organise this, or even repatriate to take care of their parents' finances. Of course, some of these fact-finding tasks are carried out during visits; we return to those later in this chapter.

Macro-factors in access to transnational communication

Access to technologies of transnational communication

As mentioned, involvement in caregiving across national borders assumes the availability of cheap and appropriate means of communication. However, there are *macro forces* at work that create obstacles to such long-distance communication. We noted in Chapter 5 that refugees incur much higher costs in making telephone calls than any of the migrants we interviewed; they lack access to a safe mail service and to email or other modern IT. Further, they cannot assume reliable and safe arrangements for the transfer of money to their relatives abroad. Researchers such as Pahl (2001b) have noted gender and class differences *within* nations in respect to access to new financial facilities. However, our findings reflect general institutional disparities *between* nations in the quality of their public facilities (such as banking, postal and telephone services), which we believe require policy resolutions at an international level. For example, in this age of global communication it would appear worth investigating why Australian providers of international telephone services charge much higher rates for telephone calls to China, Eastern Europe, Africa or the Middle East than to Western Europe, North America and New Zealand, while offering discount rates only to the latter three regions. It is also worth exploring, given the reach of major international banking institutions, why the safe electronic transfer of monies to any 'Western' nation can occur readily, while monies to some non-Western countries need to be smuggled in – with the attendant problems of illegality and the likelihood of theft.

An interesting aspect of the costs migrants incur in making telephone calls came to light in a pilot study we undertook in 1999 prior to our main project (Baldock, Baldassar & Lange 1999). This involved focus group interviews with migrants from a wide range of backgrounds, including Vietnamese migrants who had chosen to remain customers of one of the most expensive providers, even though privatisation of the telephone system had brought a plethora of other cheaper ones. Vietnamese participants explained their choice by saying that Vietnamese people felt a special loyalty to their provider, because this firm had portrayed itself as a genuine Australian company. At the same time they expressed some ill feeling about the fact that they were not being rewarded for their loyalty by means of cheap telephone calls. While the provider in question had cheap call rates to North America and Europe, it did not have them to Vietnam. The Vietnamese population in Australia may have the potential to lobby for bringing about change in this situation; this, however, they had not done.

The maintenance of intergenerational language

The above macro-factors concern access to affordable technologies of communication, but there is another important aspect to transnational communication: in order to provide emotional comfort and support, there is need for a *common* language (Ackers & Stalford, 2004). Transnational migrants usually maintain their ability to *speak*, although over time some lose the confidence to *write*, in their native language. However, unless multilingualism is supported not only in the home, but also in a variety of formal settings, including formal schooling, young children tend not to maintain their parents' language.[3] This means that elderly grandparents may well be deprived of the emotional comfort of being able to talk freely to their grandchildren abroad. Both migrants and their parents in fact raised this issue.

Australia is a multicultural, but primarily monolingual country. 'Foreign' languages are taught in High Schools, but, particularly over the past decade, there is limited space for maintenance of community languages spoken by migrants. This is notwithstanding the development in the late 1980s of a blueprint for an innovative National Policy on Languages (Boyd, 2006) with as goals – apart from the provision of English for all Australian residents – the teaching and learning of a language other than English for the purpose of 'enrichment of the cultural and intellectual life in Australia', as well as equitable and widespread language services. The wording of these goals suggests that there was a genuine interest at the time in supporting Languages Other Than English (LOTE). However, data on the implementation of the National Policy on Languages indicate that funding was directed mainly to English as a Second Language (ESL) programs for new immigrants. Some monies were also allocated for second language learning support, but whatever benefits derived from such programs were wiped out in a 1991 White Paper, in which English language and literacy were given absolute priority, and subsequently in 1994 with the formation of a National Asian Languages and Studies in Australian Schools Strategy (NALSAS). In accordance with the latter, secondary schools were asked to introduce four priority Asian languages: Japanese, Indonesian, Chinese and Korean. Significantly, as Boyd (2006) points out, these are 'foreign' languages learned in order to communicate with economic interest groups in Asia, *outside* Australia, rather than to facilitate communication with migrants within Australia. The four priority languages chosen are not prominent in the Australian community; other Asian languages which *are* important migrant languages, such as Malay, Vietnamese, Hindi, Thai, Cambodian and non-Mandarin varieties of Chinese, did not receive government support. Boyd concludes from her analysis of Australian language policy that, despite continuing immigration, the community envisaged by Australian Commonwealth policy remains monolingual with a focus on ESL and English literacy rather than on 'the complementarity of English and community languages and other LOTEs' (Boyd, 2006) that had been apparent in the policies of the 1980s. Important in this context is the fact that English is the taken-for-granted majority language, which, says Boyd (2006), 'makes it difficult, at times, to motivate monolingual Australians to learn other languages'. Boyd notes further that the multilingualism of migrants is considered of less importance than the much more limited multilingualism of those learning foreign languages. Such multilingualism, if it occurs at all, does not begin until High School; the primary school remains 'as a monolingual habitus' (Extra, 2006). In that context, there is little incentive for second-generation migrant children to maintain the community language(s) of their parents and grandparents. Speaking another language relegates them to membership of an ethnic minority, rather than strengthening their position as members of the dominant 'Australian' majority (Chiro & Smolicz, 1998).

It is instructive to compare these Australian developments in Language Policy with the new requirements for multilingualism in the European Union.[4] The language policy being developed in the EU to accommodate the coming-together of a large number of nationalities takes as starting points that (1) multilingualism is normal and (2) monolingualism exceptional (Nelde, 2006). Policy proposals imply that all children ought to be introduced to three languages: the standard language of the nation-state in which children reside; English as a *lingua franca* for international communication; and an additional third language chosen from a set of priority community languages common to major migrant groups in the region (Extra et al., 2002; Extra, 2006). The introduction of a three-language curriculum has not yet been elevated to policy and there is some resistance to its introduction in members' states (Extra et al., 2002; Boyd, 2006). Interestingly, the resistance is not caused by a monolinguistic trend towards prioritising the local majority language. Rather, what is at stake is that English, as *lingua franca*, is encroaching on the learning of the local majority language. Boyd (2006) and Extra (2006) illustrate this with examples drawn from Sweden and the Netherlands, respectively. The outcome of this is a strong push towards learning and maintaining the local majority language, be it German, Swedish, French, Dutch or Italian, because 'the interest of the state, played out in education and other forms of secondary socialisation, is to create and maintain a citizenry which is proficient in the language(s) of the state' (Boyd, 2006).[5]

What remains significant is that (with the exception of the UK) all children are at least motivated to become bilingual, receiving formal tuition in English *and* the dominant majority language in their country. This is in contrast to Australia, where the push in formal education is towards learning and maintaining the dominant majority language *only*. Given the continued efforts in Europe to implement a genuine multilingual policy with focus on 'minority living languages', as well as

English and the local dominant language, it goes without saying that children on the European continent have a greater opportunity to learn their parents' native language than children in Australia. It is worthwhile remembering in this context that so-called expatriates, who work in Australia on contract with major international industries, have the *right* to employer-paid language training for their children. In their case, then, macro-policies attempt to ensure that grandchildren remain able to communicate with transnational grandparents, whether long distance or face-to-face during visits. Other second-generation immigrant children in Australia do not have that opportunity. Learning a community language after school hours in a community language centre (as some migrant children are encouraged to do by their parents) is much less likely to lead to maintenance and promotion of that language than learning the language as part of a formal school curriculum (see also Extra, 2006).

Meso-factors in transnational communication of care

Community organisations and access to services from a distance

Meso-factors concern support given to, or barriers imposed on, transnational communication of care at community level. Support could entail a willingness on the part of community members or community organisations to assist with transfer of money or letters, as we have found in the case of refugees. Barriers to caregiving from a distance might occur when no facilities exist in the local community (either in Australia or in the home country) for migrants to gain access to information about the care their parents require. It is worth noting in this context that in a number of 'Western' countries such as the United States, Canada, the UK, the Republic of Ireland and the Netherlands, community organisations which specialise in aged care and support for carers have developed sophisticated websites from which local carers can draw information, advice and support.[6] Some of these websites cater also for translocals, long-distance caregivers within the same country. For example, the US-based Family Caregiver Alliance has published a *Handbook for Long-Distance Caregivers* (Rosenblatt & Van Steenberg, 2003) that can be accessed online,[7] while the American Association of Retired Persons (AARP), has dedicated a special section of its website to long-distance caregivers.[8] Again in the USA, the responsibilities of long-distance carers are recognised in that they are able to buy the services of an elder-care (or geriatric) care manager to make the necessary decisions on their parents' care, especially regarding suitable accommodation.[9] There are also various organisations providing service to caregivers, which are experimenting with the creation of translocal caregiving programs. For example, the Alzheimer's Association of Los Angeles has initiated a program for translocal caregivers that entails the services of a family consultant and the development of resource guides. Long distance in this instance was on average 304 miles.[10] In these cases, then, facilities have been created at the community level to assist translocal caregivers.

It is important to reiterate that long-distance translocal caregivers in the examples given above live within the same country as the recipient of care, unlike the transnational migrants in our study. But much of the information and advice provided to them, especially on the websites, could apply transnationally. For example, the *Handbook for Long-Distance Caregivers* indicates that long-distance carers can undertake extensive research by computer and telephone to assess available resources and services in the area where their care recipient lives, prior to any visiting. Transnational carers, assuming they have access to the Internet and relatively inexpensive telephone services, can also undertake such research.

We did encounter a few examples of transnational migrants searching the Internet to gain information, but formal contacts with professional care managers in their home country were uncommon. Generally, when some type of care package is required, local carers in consultation with professional caregivers develop it, while transnational migrants give advice (where sought) and provide assistance to their local caregiving siblings during visits. The transnational migrants who participated in our study, then, do not have independent access to professional care managers in their home country. This is except for a very few instances where they make incidental phone calls to their parents' doctors.[11] Not surprisingly, those transnational migrants who had medical or nursing training were most successful in their efforts to participate in the development of care packages for their parents.

The above examples (mainly from the USA) suggest that it may be possible to institute services at community level that assist transnational caregivers who need to provide care from a distance. Such initiatives could work on a regional basis, say between transnational migrants from a specific region in Italy or Ireland and representatives of community organisations in their region or town of origin. An interesting regional initiative mentioned by a Sicilian interviewee concerns a comprehensive tourist agency he started for visiting migrants. This includes picking up tourists from the airport and providing them with lodgings, researching the ancestry of migrants who want to know more about their roots, regularly placing flowers on the grave of deceased family members of migrants, looking after local houses owned by migrants and even providing some care services for elderly parents of migrants. The interviewee boasted a total of 80 different services, offered to migrants who originate from specific regions in Sicily but are now spread out over the world, mainly in North America, Australia and Canada. He operates this as a family business with help from his wife and two other family members. Although a commercial enterprise, it is very small scale and focused entirely on a specific region. It would appear quite feasible for such migrant services to be carried out by local or regional community organisations. We return to the possible role of community organisations below.

Transnational visiting to give care

As discussed in Chapter 6, visits are necessary for emotional and moral support; and to enable the participation of transnational migrants in practical and personal care. They are also important to gain information and to negotiate care arrangements with local carers. The actual decisions that have to be made regarding care arrangements require close ongoing attention and can generally only be accomplished by local caregivers or by transnational migrants during visits back home. Even then, only those who are able to make lengthy visits can take part in the formulation of a care package, which includes the choice of appropriate accommodation or negotiations with professional care managers. In order, then, that they can be involved in all aspects of hands-on, face-to-face care, transnational migrants must have the opportunity to make regular visits of appropriate duration, as well as short visits at times of crisis. This, however, is not always possible due to a number of factors.

Macro-factors in access to visits

Migrants going home

We demonstrated in Chapter 6 that transnational migrants who wish to visit their home country may encounter obstacles in doing so, for reasons including visa regulations, health insurance issues, employers' leave arrangements and community or family pressures related to finances or gender issues.

Visas are a problematic issue in the case of *refugees* wanting to visit their relatives living in transit in Iran. To gain a visa, visitors require a sponsor in Iran, the visa only applies to the one city that is specifically identified in the visa, and the process of application takes at least a month. We pointed in earlier chapters to the fact that relatives in Iran are themselves refugees, who face constant scrutiny from the Iranian authorities.

All this makes the journey of Australian-based refugees to Iran a perilous matter. Because of such difficulties, some refugees meet with their Iranian-based family members in other neighbouring countries.

On the other hand, the *migrants* who participated in our study generally have no difficulties returning to their home country. Citizens of Australia, as well as the USA, Canada and New Zealand, have privileged status in terms of access to EU countries; this is in contrast to citizens from a long list of other countries, mainly African, Asian and from the Middle East, who all require visas. Australian citizens only need a valid passport and no visas to enter Ireland or the Netherlands if they visit for family or tourist purposes; however, they will require visas if they undertake business trips. Business travel is covered under the so-called Schengen Agreement, which includes 11 European countries; Schengen visas allow a maximum stay of 90 days within any six-month period in these countries. In the case of Italy, visas are only required for visits of more than 90 days, while visitors to Singapore need a Social Visit Pass, valid for up to 90 days for people travelling as tourists. Such a Pass is issued at the airport on arrival after visitors have shown a valid passport, a return ticket and evidence of sufficient funds to finance their stay. For visits to Singapore of more than three months, a long-term Social Visit Pass is necessary; this requires a local sponsor. Interestingly, although migrants with Australian citizenship who return to European or Asian homelands have no difficulty doing so, many interviewees who have lost their citizenship say they feel very aggrieved when going through customs at airports in their country of birth that they have to line up in long queues with 'foreigners' rather than being fast-tracked with citizens of their homeland. The above issues, of course, apply only if migrants have taken out Australian citizenship and as a result lost the citizenship of their home country. Those who retain citizenship in their native country are able to visit home without even the slightest entry restriction. However, as soon as they step on native soil, they may be asked to fulfil citizenship obligations, such as military service or payment of taxes – reasons that may well make them hesitant to visit.

People without Australian citizenship who want to travel abroad require a *Resident Return Visa* (RRV) to enter Australia upon their return. For migrants who spend most of their time in Australia, a convenient option is the five-year resident return visa for permanent residents who, when they apply, must have been physically in Australia for at least two of the previous five years before lodging an application for the RRV. Those who do not satisfy that criterion may apply for a three-month return visa, which is available to people who for compelling or compassionate reasons, for example, the death of a family member overseas, have to return to their country of birth. For those who have been in Australia for less than two years, frequent trips out of the country could considerably delay the date on which they can apply for citizenship, because the Citizenship Act 1984 states that they have to have been in Australia at least two out of the last five years and a total of at least 12 months within the last two years prior to their citizenship application.[12]

Asylum seekers (that is, on shore 'illegal' refugees) do not qualify for any travel documents until they have been in Australia for two years and therefore they have to wait until they receive some legal status from Australia before they can travel. In fact, by definition, refugees cannot go back to their country of origin, at least in the short term, without having their refugee status questioned and possibly revoked. In contrast, offshore 'legitimate' refugees are given permanent resident status on arrival in Australia and they can obtain travel documents within their first two years if they require them. However, there is a perception among many that such papers do not have the same status, nor do they afford the carrier the same protection, as an Australian passport. Thus, many are unwilling to travel overseas in this early period of settlement. *Migrants* who are permanent residents may also be unclear about government regulations affecting their residency in Australia and uncertain about how long they can legitimately stay out of Australia without jeopardising their permanent residence status or application for citizenship. The uncertainty and lack of clarity that both migrants and refugees have about visas and their rights of re-entry to Australia are an added stress with which those involved in transnational caregiving have to contend.[13] In this context it is important to remember that 'passports are necessary *and sufficient* not for gaining entry to another country but only for returning to one's country of origin' (Torpey, 2000, p. 164, original emphasis).

In addition to proper travel documents, long-distance travel generally presumes some form of *health and travel insurance*. As we mentioned earlier, Australia has Reciprocal Health Care Agreements with a number of countries: Finland, Italy, Malta, the Netherlands, New Zealand, Norway, Sweden, the United Kingdom and the Republic of Ireland. This means that upon displaying a valid Medicare Card in these countries, Australian citizens and permanent residents have access to basic health care – the details of which vary from country to country, but which is usually limited to inpatient hospital care.[14] This arrangement is helpful to transnational migrants, although it is quite clear that many Australian residents are entirely unaware of this entitlement. Even in the case of travel to countries with Reciprocal Health Care Agreements, the Australian government recommends that anyone who travels abroad should take out private health insurance. This can be a costly matter, with full coverage for a family priced at hundreds and sometimes thousands of dollars. Furthermore, most policies are not available to people over the age of 70. This is a problematic issue for migrants over the age of 70 who want to visit their country of birth, sometimes even then for caregiving purposes as care to siblings or even parents may continue. Australia does not have a reciprocal health care agreement with Singapore.

The above issues apply to routine visits, which can be planned and saved for ahead of time. Paid workers who take such routine trips are able to travel when their employers allow them to take their annual leave, or long-service leave. The *Workplace Relations Act 1996* stipulates that, in Australia, employees who are not casuals are entitled to a minimum of four weeks paid recreation leave for every 12 months worked. An additional long-service leave of 13 weeks is normally granted after 10 years of continuous service. However, some employees might be limited by conditions restricting the times of year when they are able to take this leave.

When a sudden crisis occurs, visas, airline tickets, travel and health insurance, leave from paid work and airfares must be found at one or two days' notice. While airlines, travel agents and insurers are generally willing to act quickly (and have procedures in place to expedite matters in such crisis situations), employers are not always helpful. This is especially difficult when transnational migrants need to attend a funeral in their home country. The *Australian Workplace Relations Act 1996* guarantees two days of paid bereavement leave on the occasion of the death of a close family member.[15] This is totally inadequate for employees who need to travel to Europe: travel in one direction alone takes about 24 hours. It is worth noting in this context that many transnational migrants keep a substantial sum of money in a special account, so they have funds to pay for airfares in crisis situations.[16] We return to the issue of finance for air travel when we consider meso-factors in access to visits.

Parents' visits

So far we have focused on travel by migrants for return visits home, but parents, of course, also travel to receive or provide care. We found that it is common for Dutch, Italian and Singaporean parents to visit in order to provide care after the birth of a child, while Singaporean parents may visit for substantial periods so that their migrant children can care for them. Whatever the purpose of their visits, any parents travelling to Australia are required by the Australian Immigration authorities to apply for a visitors' visa. Such visas are granted only to 'genuine' visitors who have come as tourists or to visit an Australian citizen or permanent resident. To be deemed genuine, they generally need to show evidence of a return ticket and sufficient funds – in their own bank account or that of the Australian resident they are visiting. There is, however, more to the notion of 'genuine' than first meets the eye.

There are in fact three categories of visa applicants (Burn & Reich, 2005, Chapter 19):

(1) *Desirable applicants* from a list of 33 countries, who are clearly very welcome in the eyes of the Australian immigration authorities, as they represent high volume and low-risk tourist trade. These prospective visitors can apply for an Electronic Travel Authority (ETA) through their travel agent, or even at the airport; they pay no fees and fill out no forms. They are not asked questions about their health status, and their visa may allow multiple entries over a period of 12 months, up to a maximum of three months each time. The list of 33 countries includes the home countries of all migrants who participated in our study. In other words, parents from Italy, Ireland, Singapore and the Netherlands may apply for these easily accessible visas. New Zealanders, of course, do not need them because they enter Australia freely by just showing their New Zealand passport.

(2) *Regular applicants* are visitors from any other countries, with the exception of those listed as high risk (see below), and they need to apply for a regular visitors' visa that will be stamped in their passport, allowing multiple entry, usually over a period of four years, and entitlement to stay for three months (short stay) or six months (long stay) at one time. Interestingly, parents aged between 60 and 70 years may be given a visa for twelve months. Members of a family travelling together (such as a husband and wife, and dependent children) must all apply separately and each pay a fee of $65.[17] Like visitors on ETAs, they must prove they are genuine, with return tickets and evidence of sufficient funds to cover their costs during the visit. These applicants will be asked questions about their health status, and if they are older than 70 they are expected to carry a 'fitness for travel certificate' from a doctor and show that they are financially capable of meeting the cost of any medical treatment. This means they must have sufficient funds, or take out private health and travel insurance, or alternatively be eligible under a reciprocal health care arrangement between

Australia and their home country. Such Reciprocal Health Care Agreements provide coverage either for the duration of the visit or for a total of six months, depending on the country of origin. Reciprocal health care arrangements offer some health care to any parents visiting from Ireland, Italy, New Zealand and the Netherlands, but not to visitors from Singapore.

Visitors (including those on ETAs) who are ineligible for Medicare benefits, or choose to be treated as private rather than public patients, are encouraged by the Australian government to take out private health cover. Again, this can be very costly, especially for those over the age of 70.

(3) *High-risk applicants* are potential visitors from countries that are deemed to be 'high risk'. The notion of high risk is based on a profile developed by the Department of Immigration of visitors who are likely to overstay their visitors' visa, or 'contrary to their stated intention, apply for a permanent visa once they are in Australia' (Burn & Reich, 2005, p. 653). The list is highly specific, taking into account factors such as nationality, marital status, age, sex and occupation. Australia's high-risk list is similar to lists maintained in EU countries and in the USA; people from African, Asian and Middle-Eastern countries are singled out for specific scrutiny in all cases while there are no applicants from Western Europe or North America on such lists. This means that the Australian list does not include the United Kingdom, which according to Australian Immigration statistics produces *more* people who overstay visitors' visas each year than any other country.

The latest version of the Australian list includes four categories of applicants who appeared in our study: females, 60 years or over from Afghanistan; males 25 years and older and females 30 years and older from Iran; and females, 60 years or older from Iraq. People singled out in this manner must give evidence of a higher level of genuineness than other applicants. If they fear they will not be successful, they may apply as 'sponsored' family members for a visa normally valid for three months only, requiring an Australian resident as sponsor who will provide a refundable security bond, usually in the range of $5000–$15,000 (Aus). As of April 2005 people on sponsored visitor visas may stay up to 12 months. However they cannot receive a multiple entry visa, and need to go through the entire cumbersome application process again for any subsequent visits. Even if granted a regular visa (rather than a sponsored one with the added costs), they cannot stay longer than six months.

Of the different sample groups, Singaporeans had the most to say about problems with visas and health insurance. They noted the costs of visas, and the paperwork involved, saying that application for a visa beyond a period of three months is costly, time-consuming and complicated, as it involves a medical examination including X-rays, the cost of which one person remarked, 'is higher than an air ticket back to Singapore'.[18] Singaporean migrants and their parents were particularly dissatisfied with the fact they are required to pay such large sums of money for health insurance. In one case, a migrant woman pays for her mother's medical insurance when she visits from Singapore, but this is not necessary in the case of her mother-in-law who visits from the United Kingdom. A policy change in this regard would appear in order. Given the high number of Singaporean migrants in Australia, and the major economic contributions they make to Australia from their student days onwards, it appears a strange anomaly that Australia and Singapore have not negotiated a Reciprocal Health Care Arrangement as has been done between Australia and other close migration partners. Singaporean interviewees themselves offered various policy solutions, such as 'some special insurance rate for visitors who have family here'.[19]

The various obstacles mentioned above are due to Australian visa and health care regulations. There may be other factors at play in their home countries that restrict the ability of parents to visit for long periods. This was clear in the case of one Singaporean mother who said: 'I cannot go to other countries for more than three months. The government will stop paying me my Singaporean allowance [on top of my pension] if I do'.[20]

Meso-factors in access to visits

Transnational migrants with limited financial resources will find it difficult to pay for the expense of air travel, especially if they have to make a return visit at short notice because of family crisis or bereavement – long-distance airfares from Australia are invariably much more expensive without advance purchase. We have described in Chapter 4 to what extent refugees can call upon the financial support of members of their refugee community for such expenses, as a form of generalised reciprocity between individual families. We have found only one instance of a more formal, community-based support system among migrants. This had been developed within the Irish community of Western Australia, where regular community activities such as dances and raffles generate funds that among other things can be made available in the form of loans to Irish migrants who have to travel at short notice for a family crisis.

Finances are not the only concerns migrants have when they need to make a crisis visit. If a migrant woman is returning home by herself (as happens often in a family crisis concerning her parents) she will need some assurance that her partner and children receive support while she is away. Of course, in some cases such help may come from other relatives, friends or neighbours. As with finances, only in the case of the Irish is such support, especially for the provision of childcare, part of a more formal community arrangement. The organisation of a community support system of this kind, for the provision of loans, childcare and possibly even to look after migrants' homes while they are overseas, would not be particularly difficult to arrange. Most migrant communities have active social clubs and welfare organisations that could take such arrangements on board. The existence of such support structures, if well publicised in the community-at-large, may also help to dispel some quite negative reactions transnational migrants sometimes receive from fellow community members who have not experienced the traumas of transnational caregiving. We recall, for example, the case of a migrant couple who were asked to make a considerable financial contribution to their local church, and who in that context were told that they clearly had the money to spend on expensive holidays abroad (and therefore should have some money to spare for their church as well). The remarks upset these migrants, as their regular trips abroad are *not* meant to be holidays, but rather what they consider obligatory caregiving visits to elderly parents.

We have documented in earlier chapters that during their visits back home the majority of transnational migrants receive financial support, accommodation and any other assistance from their immediate family. At the same time, there is some support at a broader community level. We heard, for example, from Italian migrants that their hometowns had community facilities for their accommodation. It is also worth mentioning that in recent years travel agents with close ties to migrant communities have arranged for migrants' group travel to their home country – the Sicilian case we discussed above is not the only example. People participating in this kind of group travel are often older migrants who prefer not to have to make their own travel arrangements. They may not travel for caregiving purposes, but they will take advantage nonetheless of the opportunity for easy travel to revisit their homelands and families. In Italy, many provincial migrant associations organise annual tours to Australia to attend reunions hosted by immigrant associations across the country. Many Italians avail themselves of this opportunity to visit their Australian migrant kin in what they see as safe and familiar company for travel that would otherwise be considered too difficult to negotiate due to lack of English language ability.

Aged migration

An alternative form of caring for aged parents is to bring them to Australia. Transnational caregivers are only able to participate in practical or personal care during visits. They may well find during such visits that their parents do not receive adequate care, perhaps because there are no local siblings to give care, or due to lack of professional caregiving facilities. It is in this kind of situation that transnational migrants may contemplate whether it is possible to bring their parents out to Australia, so they can look after them on a day-to-day basis. We mentioned that, if such aged migration is not possible, in some instances migrants repatriate for the same reason.

Macro-factors in aged migration

The process of bringing out parents to Australia is fraught with difficulties. One of the main hindrances is the Immigration Department's restricted notion of family, which generally excludes parents from the definition of 'members of the family unit' which, apart from the primary applicant (the 'family head'), defines as secondary applicants the spouse, dependent children and other dependent relatives (Burn & Reich, 2005, p. 88). For a parent to be included in this latter category, they must meet the usual health, character and public interest criteria and also demonstrate that they are single, usually live in the household of the primary applicant and are financially dependent upon that applicant; 'other kinds of dependency such as psychological or social dependency are not considered' (Burn & Reich, 2005, p. 251).

The main avenue for sponsoring aged parents, that is, 'a parent who is old enough to be granted an age pension under the Social Security Act 1999' (older than 60–65 years of age), by an Australian resident or citizen is under the Family Stream of migration.[21] Fees for this visa include a $1245 first instalment and $1110 second instalment (Burn & Reich, 2005, p. 227). In 2003 a further category of 'Contributory Parent' visas was introduced for those parents willing and able to pay a significantly higher 'second instalment' fee – of up to $26,475 in 2004 (Burn & Reich, 2005, pp. 233). In addition, sponsors must arrange payment of a refundable social security bond of $10,000 for the primary applicant and $4000 for secondary applicants over the age of 18 (Burn & Reich, 2005, pp. 234–237). Quotas are placed on all categories of parent visas, with clear preference for the latter class: in 2003–2004, 1000 Parent visas (subclasses 103 and 804) were granted, and 5500 Contributory Parent visas (Burn & Reich, 2005, p. 11). In other words, it is easier for parents to enter Australia if they or their migrant children are wealthy.

A further requirement for aged parents is the 'Balance of Family' test, meaning that applicants have at least an equal number of their children living lawfully and permanently in Australia as in their home country, or more in Australia than in all other countries combined. The stated purpose of the Balance of Family test is to establish the nature of a parent's ties to Australia and the support likely to be available after arrival in Australia. The *quality* of the relationship between migrants and parents is said not to be relevant in determining parents' eligibility for migration and the rule is applied strictly, without any flexibility (Burn & Reich, 2005, p. 230). In other words, the Balance of Family test does not take account of negotiated family commitments or 'favourite children' (see also Aldous, Klaus & Klein, 1985). Our study shows clearly that cultural obligations may also determine which person is the most appropriate caregiver.

As mentioned, the number of visas issued in the Family Stream is capped (thus, a quota is set each year) which means that once the planning level for the visa class has been reached, no further visas are granted in that financial year. We noted in Chapter 2 that there has been an overall reduction in the number of admissions under the Family Stream in recent years, while admissions under the Skilled Stream have increased. The reduction affects aged migration more than other categories in the Family Stream, as 'the law prevents the minister from putting a limit on the number of visas granted on the grounds of being a spouse or a dependent child' (Burn & Reich, 2005, p. 13), although a quota may be placed on prospective marriage visas.

We have mentioned already that there is an additional barrier to aged parent migration and visits in the fact that certain countries are regarded as 'high risk' and aged parents from those countries have difficulty obtaining any visas, even for short stays in Australia. If ageing parents manage to fulfil the various criteria required of them and they migrate, the financial obligations for their migrant children can be quite prohibitive. Any social security payments made to the parent within two years of migration will accrue as a debt for the sponsor in accordance with the 'assurance of support' signed during the migration application process. No person is eligible for an age pension until they have been a permanent resident in Australia for at least 10 years (Burn & Reich, 2005, p. 505). The sponsor is expected to provide not only day-to-day living expenses but also pay for any medical or hospital expenses their parents might incur.

One of the key assumptions behind restrictions placed on the migration of parents to Australia is the predominant view held by governments, including in Australia, that the elderly are a 'burden' on welfare systems and that caring resources flow only one way, from migrant or government to the ageing parent, rather than the reverse. This contradicts a growing body of research that indicates that aged parents make significant contributions to their families and communities (see Ackers & Dwyer, 2002; Grundy, 2005). The notion of burdensome elderly is related to a tendency to view old age as a uniform stage in the life course, characterised by increasing dependence, with no consideration for the financial, emotional and practical support such ageing migrants might bring to their migrant children and grandchildren. As to the financial contribution aged migrants make to Australia, many parents on government pensions from other countries are able to bring these with them;[22] self-funded retirees who enter on Contributory Parent visas also bring substantial sums of money across.

It is quite instructive to compare these very restrictive and costly immigration laws for older migrants with the free and easy movement of aged persons within the context of the European Union. Many elderly Europeans from Northern countries migrate south to Spain, Portugal or Italy upon retirement, motivated by their desire for a better climate, increased opportunity for travel and the removal of legal and institutional barriers to freedom of movement after the Single European Act of 1986 (King, Warnes & Williams, 2000, p. 31). For example, 'translocal' migrants are now able to acquire property rights in other EU countries, and are able to cast their votes in local and European parliament elections wherever they live (p. 32). King and colleagues (2000, pp. 175–178) observe that there are still 'marked discrepancies among the EU countries in the formal provisions of social security payments, public-sector medicine, and specific measures for older and disabled people'. At the same time, there has been considerable progress in the direction of unification of citizenship rights. Especially important in this context is the notion of 'European Union citizenship', conferred on everyone holding the nationality of a member state. The rights accorded to EU citizenship include freedom of movement for residence. In the case of retired people, this applies to those 'who have a pension and health insurance or sufficient resources to prevent their becoming a charge on the host country' (King, Warnes & Williams, 2000, p. 175). Residence permits granted in this context are valid for five years and renewable and the cost of such translocal migrations is insubstantial compared with transnational aged migration to Australia, because in the European case there are no costly visas or other fees to be paid.

Repatriation

The methodology used in our study did not encompass a focus on transnational migrants who have repatriated. Nonetheless, we encountered several cases of repatriation, in some instances for the sake of giving care to elderly parents. A permanent or long-term return to the homeland for the purpose of caregiving was most likely in the case of Italian migrants because of their strongly held belief that they should participate in their parents' care. Whatever the reason for repatriation, a variety of obstacles need to be overcome in the process. Crucial, of course, is the right of re-entry to the country of birth: only those who have retained their nationality or have dual citizenship will be able to move freely, without the need to re-migrate. Many interviewees in fact noted the possibility of permanent or long-term return to their country of birth as a reason for not applying for Australian citizenship. This was especially relevant to Dutch migrants who knew they would lose their Dutch nationality if they took up Australian citizenship. The Dutch government has changed this rule only very recently – as, indeed, has the Australian government. Prior to 2002, the Australian government did not allow Australian citizens who took up citizenship elsewhere to retain their Australian nationality (Burn & Reich, 2005, p. 795).

Of great importance is also the portability of skills. Many migrants – for example, doctors and nurses – find it difficult to get recognition in Australia for their professional qualifications, often having to retrain and requalify. Repatriation brings similar problems. Some migrants we interviewed return regularly to their country of birth to update their qualifications and maintain registration so as to ensure their skills will be portable if and when they decide to repatriate. Migrants, returning back home with children, face the additional difficulties of their children's adjustment to different school systems and possibly lack of facility with their parents' native language. The cost of moving is another important consideration.

As we mentioned in Chapter 3, the only group of migrants in our study that seems prone – and faces only limited obstacles – to long term but not necessarily permanent return are the *Irish*. Prime factors that ease the process of repatriation in their case include ease of re-entry (due to the availability of dual citizenship), similarity in culture and language between Australia and Ireland, and support from their families in obtaining jobs and accommodation. Importantly, Australian-born children do not face language problems when settling in their parents' home country. Indeed, a number of Irish migrants experience this ease of repatriation as a strong pressure to return.

It is worth noting that sometimes people who have come to Australia as migrants want to return to their homeland in their old age, leaving their adult children behind. Some of these are people who begin to suffer from homesickness as they age and fear the prospect of being buried in 'foreign' soil. Those who have been in Australia for ten years or more are eligible for the Australian aged pension when they repatriate (Social Security Act, 1991). Nonetheless, the move back home of such elderly people can place added burdens, both emotionally and financially, on their Australia-based children. In fact, such repatriations start a new cycle of transnational caregiving.

Community organisations and micro-factors in transnational caregiving

Negotiating tensions between caregivers

It is reasonably well understood today that caregiving is a difficult task, and that tensions may arise, either between caregivers and the people they care for, or between primary caregivers and others who take some responsibility for caregiving. Such problems as exist have led to the creation of self-help carers' organisations, where carers give each other mutual support, and some form of counselling is available from volunteer or paid staff. It is not always recognised that transnational migrants also experience tensions in their caregiving tasks. Such tensions take several forms. Transnational migrants may experience stress and anxiety by the sheer fact that they live so far away from the persons they care about, without adequate opportunity to help out or visit. They may face conflicts with their spouses and children, for example, over the costs of sending financial support to parents (as in the case of refugees), or the costs of travel to visit parents. Further, they may experience conflicts with local siblings over issues such as the type of care needed, or the degree of access of transnational migrants to their parents, if these are, for example, cared for in the home of a local sibling. They may experience severe stress over decisions as to whether to apply for visas so their parents can migrate to Australia (and the consequences of such migration for their own family) or to repatriate.

We reported earlier on an American experiment in service provision to long-distance caregivers of people with Alzheimer's. One of the most interesting findings of the research that accompanied that project was that long-distance (translocal) caregivers frequently experienced conflicts with other family members. For example, when a patient with Alzheimer's was being cared for by a spouse, this spouse would often maintain that s/he did not need any help with caregiving, against the better judgments of the professional family consultant or the long-distance daughter or son who had called for professional advice. Similarly, some siblings who lived locally would resent an incursion on their sense of control over the caregiving process, when their long-distance siblings wanted to arrange a care package with professional advice. Such examples are very much in line with our own findings that transnational migrants who are caregivers face considerable conflict and stress, which requires resolution. What can be done to assist transnational migrants in this regard?

Earlier in this chapter we raised the possibility of community organisations providing assistance to transnational caregivers in gaining information relevant to caring from a distance. In our view, such community organisations could also be galvanised with financial assistance from government to provide counselling to transnational migrants, and to create self-help groups of transnational caregivers to help deal with the anxiety and anguish of caring across borders. In other words, we suggest the possibility of establishing associations of *transnational* carers along the same lines as carers' associations for *local* carers. It may be feasible to incorporate Internet chat rooms for transnational carers in such programs. We are certain that transnational migrants who experience conflicts with spouses, or with siblings back home, would find such support groups helpful in clarifying issues and coping with stress.

The *resolution* of such conflicts is a bigger problem, which will require more specific action. A first requirement is that professional caregivers such as doctors, social workers and aged-care providers in the *home country* are educated to recognise the importance of transnational caregivers in the caregiving process. They have to learn that the absence of cherished migrant children may have an ill effect on elderly people, that most transnational migrants seek an involvement in decisions made about their parents' wellbeing and that they should be able to access information about their parents' state of health, so that they can assist in such decision-making. In other words, the invisibility of transnational family members needs to end. At the same time, professional service providers in the *host country* need to have greater knowledge about the burdens of responsibility many transnational migrants face, ranging from financial difficulties due to costly visits and overseas remittances, to emotional trauma and depression caused by lengthy separation from loved ones and anxiety about their well-being. In order to deal with persistent stress between transnational and local siblings, professional service providers may facilitate the creation of a 'team' of family members both in the home and host country to decide together what needs to be done for a particular elderly family member. Conference telephone calls or perhaps even transnational family chat rooms could be arranged for such purposes.

Negotiating the tensions of aged migration

The migration of elderly parents to Australia so that they can be cared for, brings problems of its own which are not always recognised ahead of time by their migrant children. Parents may feel uprooted because they have lost long-standing friends and neighbours and they may feel especially isolated if they are not able to speak the language of their host country. This may also mean they cannot communicate properly with their own grandchildren. In fact, many parents of migrants we interviewed specifically commented they would never want to migrate to Australia, because it would mean losing the support of their neighbours and local community.

When these aged migrants reach the time that they become dependent on professional aged care, those who lack English language skills face special problems, similar to those of NESB migrants who came to Australia many years ago and did not have adequate English language skills (MacKinnon, 1998). The current government approach to aged-care services in Australia, to maintain older people in their own homes for as long as possible by providing home and community-based services, is considerably more difficult to implement when the older person requiring care has difficulties with the English language. MacKinnon's research on older Italian-Australians identified English language competence as being a prerequisite for effectively accessing health-related information and services, and, in particular, the ability of older people to participate in the planning and implementation of their health care. Lack of facility with the English language exacerbates the difficulties associated with cultural upheaval and discontinuity. As a consequence, many older migrants who lack or have lost English language skills are unable to do the things many Anglo-Australians take for granted, such as negotiating public transport, accessing information and the mainstream media, telephoning emergency services or lodging complaints about defective health services (MacKinnon & Nelli, 1996).

Given the many difficulties experienced by aged parents who migrate to Australia, it is imperative that they are welcomed upon arrival by a local community to provide them with companionship and support. Community organisations, again, could play a major role in this regard. There is currently no adequate solution to the problems such aged migrants will come across once they require aged care and health services. Some ethnic-specific aged care is available in Australia, through in-home domiciliary care providers who speak the native language of the ageing client, as well as through institutionalisation in ethnic-specific hostels or nursing homes.[23] However, the funding for such ethnic-specific services is inadequate and even NESB migrants who have lived in Australia their entire adult life (having migrated in the 1950s) find it difficult to gain access to the resources and networks they need in their old age. Recently arrived aged migrants who have limited community resources and networks would probably find such access even more difficult to acquire.

Dual citizenship and transnational caregiving

For most Australian-born, citizenship is a birthright that is taken for granted. It may gain some relevance for those who obtain an Australian passport as they travel overseas, or for people attending early-dawn ANZAC ceremonies as they dwell for a moment on the patriotic duties of Australian citizens at war. Whether Australian-born are consciously aware of their citizenship obligations when they vote, or their citizenship rights when they apply to join the public service, is debatable. For immigrants, however, the situation is clearly different. For them, citizenship is not a birthright, but the outcome of a deliberate choice. Some have the option of dual citizenship, and are thus able to retain formal citizenship ties with their country of birth. Others, who take out Australian citizenship, have to surrender their citizenship connections with their home country because the government in their country of birth demands this. Finally, there are in Australia large numbers of immigrants who have never taken out citizenship, and thus remain permanent residents. Census statistics for 2001 (ABS, 2005, p. 145) show that the citizenship rate is lowest for immigrants from English speaking countries such as the United Kingdom (65.6 per cent) and New Zealand (45.3 per cent) and highest for those from Asian countries such as the Philippines (92.1 per cent) and Vietnam (91.5 per cent) and non-English speaking countries such as Greece (97.1 per cent). Generally, non-citizens do not have voting rights (Commonwealth

Electoral Act 1918, section 93), and non-citizens or people who are citizens of a foreign power are not able to become members of Parliament or permanent public servants in the Commonwealth government (Commonwealth Constitution, Section 44(i)).[24]

In public discourse the denial of rights to non-citizens or people with dual citizenship has generally hinged on questions of nationalism, patriotism and loyalty. It is assumed that people who retain their original citizenship cannot have a sense of loyalty to their new country. This is expressed, for example, in the *National Agenda for a Multicultural Australia* (Commonwealth of Australia, 1989, p. vii), which states that 'multicultural policies are based upon the premise that all Australians should have an overriding and unifying commitment to Australia, to its interests and future first and foremost'. Immigrants have at times fiercely denied the likelihood of divided loyalties due to dual citizenship or non-citizenship.

As Skrbis's (1999) work among Croatians and Slovenians in Australia has shown, it is not uncommon for migrants to retain loyalty to their home country in tandem with being a resident or a citizen of Australia. Social scientists have championed such rights through concepts such as multicultural citizenship (Klymlicka, 1995; Castles, 1997), transnational citizenship (Bauböck, 1994) and cultural pluralism (Jayasuriya, 1990). Their accounts have generally focused on the public domain of political allegiances and the consequences of such for national unity. We argue, on the other hand, that the desire migrants have to retain their original citizenship may be associated with the *private* domain and more to do with the maintenance of extended family connections and an emotional attachment to one's homeland than the public domain of nationalism and patriotism.

Our data strongly support this argument. When asked questions about citizenship and national identity, migrants who have deliberately retained their native citizenship rather than becoming naturalised Australians defend their decision in emotional terms, invoking their personal sense of identity and their family relations. This applied to the majority of Dutch migrants who knew they had to give up their Dutch passport to become Australian. They made comments such as 'in my heart I will always stay Dutch'.[25] One interviewee made a moving statement about her emotional attachment to her Dutch identity when she said: 'your passport is like the key to your house; the key to my mother's house is still on my key ring – I don't need it, but I'm not taking it off'.[26]

In some cases, these migrants added that their family overseas would not like them to give up their Dutch nationality. Family members had said, 'don't do it'; others had 'been very upset about it' [the likelihood of the migrant loosing Dutch citizenship]. Keeping the Dutch passport also keeps the door open for any children born in Australia to gain Dutch citizenship so they can return to their parents' country of birth as adults. In a few instances, practical reasons were also mentioned: the opportunity to return to the home country to work there; pension rights; and easy access to travel within the European Union. Interviewees often accompanied such comments with statements indicating that they did not see the need to become Australian, because they had no interest in politics: 'I can't vote, but I am not interested in politics', 'there is no advantage to becoming Australian, I don't have political ambition'; and, 'we don't have a problem with not being able to vote' were common responses.

At the same time, all Dutch migrants we interviewed would prefer to have *dual citizenship*, because this literally gives them the best of both worlds, the continuation of emotional bonds with the homeland, as well as the opportunity to fulfil the Australian citizenship obligation of voting in federal and state elections. None of these migrants exercise their right to vote in Dutch elections. Remaining Dutch, then, had little to do with patriotism and political participation in their country of birth; it was a private issue of personal identity and family roots. One of the very few Dutch migrants who had taken up Australian citizenship had this to say:

> *My mother-in-law was very upset because we had become Australians. She said, now I don't have a son anymore. We had to laugh about that of course ... They think it is treason, you know, but it is only a piece of paper, because inside you are still the same person, inside you are Dutch.*[27]

In the few years since the interviews took place, the government of the Netherlands has opened the door to dual citizenship. It may be expected that most Dutch migrants who had until then resisted naturalisation would have embraced the possibility of dual citizenship since then. This, in fact, has happened to the majority of Irish and Italian migrants and to quite a few of the Singaporeans, but not to the New Zealanders.

The Irish migrants we interviewed were, if anything, the most eager to apply for Australian citizenship, and quite political about it. They 'never had a doubt', said they were 'committed to being Australian', and 'it was the right thing to do'. The right and obligation to vote in

Australia figures strongly in their motivations. All but one had become citizens as soon as possible. For those who had migrated before 1984, that often meant as soon as the Australian government in 1984 dropped the requirement that new citizens swear an oath of allegiance to the Queen of England in their citizenship ceremony. Others, who came later, became citizens immediately after the obligatory two years of permanent residency. Several interviewees described at length how they use both passports. For example, one Irishman said:

> *It is nice to have a second passport, to be honest. It is easier traveling. You come into Ireland on your Irish passport and into Australia on your Australian one. [Once] I made a mistake. I went back and ... stood with Africans and Moroccans for the queue in non-EU countries, and it was just the length of that queue ... just because I took the wrong passport out of my bag. [Otherwise] I would have walked straight through.*[28]

At the same time it is interesting that, unlike the Dutch, Irish migrants do not invoke a sense of national identity in their choice of dual citizenship. Apart from one, who still identified strongly with Ireland and refused to take up Australian citizenship, their references to their ongoing Irish citizenship are all about convenience (the ease of travel for themselves and their children) and not about emotional attachment to their home country. Some, in fact, are not even quite certain whether their Irish passport is still valid and do not seem concerned about travelling on an expired one. It is possible that the Irish take dual citizenship for granted, because as members of the British Commonwealth they have *always* had the right to dual citizenship with any other Commonwealth country and their sense of Irishness goes without saying: it does not need to be explained.[29]

Italian migrants gained the right to dual citizenship in 1992, and many of our Italian interviewees have taken advantage of that opportunity or said they would if it were possible – indicating by the latter comment that they are not entirely aware of their rights. The reasons they gave (flexible travel, voting rights in Australia and their children's rights to an Italian passport) are similar to those of the Irish but not expressed with the same alacrity. But yet again, there was little reference to the value of Italian identity or an emotional attachment to the home country as reasons for maintaining Italian citizenship. The Singaporean situation is a complex one, because it appears that Singaporean law prohibits dual citizenship, but it is widely understood that the government does not necessarily enforce the prohibition. Most Singaporean migrants we interviewed had permanent residency in Australia, and were thereby able to travel between the two countries easily. Some spoke about citizenship, but said they 'did not get round to applying for citizenship yet'.

Only three of the New Zealand migrants had applied for Australian citizenship, even though like the Irish, they have an unassailable right to dual citizenship. Some said they may one day, but never get around to dealing with the paperwork. New Zealanders do not need to apply for a Resident Return visa to enter Australia, and they hardly ever need a visa to any country when travelling on their New Zealand passport. One of the few interviewees with dual citizenship said he always travels on his New Zealand passport for this reason: it is better than an Australian one. Interestingly, several New Zealand migrants also expressed a sense of emotional attachment to their home country. They said 'they still called New Zealand home' and 'I'm loath to give up my New Zealand passport, I see it as a little piece of New Zealand I want to keep'.

The above examples show that there are interesting nuances in the views migrants hold about citizenship. When they know they have to give up the national identity of their home country in order to become Australian citizens (as was the case until recently with the Dutch), they often simply refuse, and justify this by invoking strong emotional attachments to their homelands (as well as a lack of interest in politics). When it is possible and useful to have dual citizenship (the Irish and Italians), migrants will take up Australian citizenship, and do not see the need to invoke strong emotions about their home country. When dual citizenship is available, but of little practical value (as in the case of the New Zealanders), not applying for Australian citizenship is again justified with reference to strong emotional attachments to the home country.

At the same time, the overwhelming majority of the migrants who have remained permanent residents (some for 20 years or more), rather than becoming Australian citizens, appear totally secure in the knowledge that they will never lose the right of access to Australia. Of course, they do experience some bureaucratic nuisance in having to apply for RVVs (Resident Return Visas), but they did not express any fear of possible changes in immigration rules that might herald loss of rights or even deportation.[30] This sense of security is interesting, because in recent years the Australian government has in fact deported some permanent residents and even citizens. In 2005 the press reported two notorious incidents: one of a German-born Australian citizen with mental health problems who was placed in a refugee detention centre; the other a Philippines-born Australian citizen deported after what appeared a traffic accident that left her incapacitated. Both explained to immigration authorities that they were citizens, but did not carry their Australian passports and were not given an opportunity to seek legal advice or notify family. Currently, a Public Inquiry is investigating at least 200 cases of deportation that may in fact have been unlawful. Possibly, migrants in our study felt secure because except in the case of deportations due to criminal convictions, few of the deportees have been 'white' and 'Western'. Indeed, critics of Australian immigration authorities have argued implied racism in such deportations.[31]

In contrast to the migrants, the refugees in our study expressed a strong desire to attain Australian citizenship and some were fearful that their applications would be rejected. For them, gaining an Australian passport and Australian citizenship is simply essential. Without an Australian passport, they believe they are not able to travel to visit their families. As they explained, they cannot go home, and their relatives cannot visit Australia, but with an Australian passport they can at least see each other in a third country. Without a passport, the chance of never seeing your family again is always there.

Whatever the nuances in views and motives about Australian citizenship and dual nationality expressed by migrants and refugees, we argue that they are related mainly to the private and domestic sphere of families and individuals. They are about ease of travel, about avoiding the paperwork needed for visas, about maintaining a sense of one's roots and visiting family. Specifically, it is important to recognise that they are about the ability of transnational migrants to fulfil their obligations as caregivers for their elderly parents who live overseas. As we have indicated above, one of the most significant aspects of transnational caregiving is the return visit back home. Citizenship status has an important bearing on the ease with which a person can visit their country of origin and return to Australia.

Because citizens are entitled to enter their country of citizenship without a visa, having dual citizenship remains the best way to ensure that migrants involved in transnational caregiving have easy access to both their parents' country and Australia. The Commonwealth government acknowledges this link between migrants' home and host countries. For example, the Australian Citizenship Council (1999, p. 14), in its paper *Contemporary Australian Citizenship* states that the settlement experiences of migrants take time and that they should not be expected to instantly shed their feelings and associations with their country of origin. However, the length of time over which migrants retain connections with their home country, the complexities inherent in those connections and the associated implications for citizenship are not explored in any detail. The existence of increasing numbers of lifelong communities of care between home and host countries suggests a need for a widening of the notion of citizenship so that it caters more equitably for the diverse citizenry which make up present day Australia.

Concluding comments

It is Australian government practice to encourage immigrants to become citizens. In doing so, it increases the need to develop policies that cater for the diverse circumstances in which migrants find themselves. Transnational caregiving has received little recognition as an area of policy development, yet those migrants who are involved in caring for overseas-based parents are affected, often negatively, by legislation and policy in a range of portfolio areas including citizenship, immigration, social security and workplace relations. A move by the Commonwealth government to a multicultural notion of citizenship that recognises the specific needs of particular communities would both take account of the difficulties many migrants face in caring for parents not resident in Australia and contribute to more equitable citizenship.

While the above issues necessitate institutional changes at the macro level of government and industry, transnational migrants also face problems at the micro level of family and community. We have argued that these can best be dealt with through the creation of transnational carers' organisations, initiated by existing community groups, with government support. Such organisations could develop along similar lines to local carers' organisations, but would benefit from advances in new technology to create transnational Internet chat rooms and other web-based resources.

8
Conclusion: Towards a Model of Transnational Caregiving

In many ways, transnational family caregiving is no different to local and translocal caregiving, in that it is characterised by a pattern of generalised reciprocity.[1] Care is exchanged between and across the generations, with the type, timing, direction, recipient, provider and flow of care changing over the life course. Whether near or far, family caregiving is informed by gender, culture, class and context and is mediated by macro- (state provision of care services), meso- (community involvement) and micro- (individual/family) factors.

This said, geographic distance, national borders and the processes of transnational migration do impact on family caregiving exchanges in a number of important ways, including which types of care are exchanged, how families go about providing care, as well as when, why and how it is practised. By way of conclusion, this chapter brings together our empirical findings in an effort to formulate a model of transnational caregiving. The objective of this model is to capture something of the processes and key dimensions involved in the exchange of care between family members across national borders. We end the chapter with a discussion of our key findings and research contributions.

In developing a model of transnational family caregiving we first need to highlight the unique dimensions of this practice of care exchange. Most importantly, transnational caregiving is characterised by the crossing of national borders and the maintenance of relationships in two (or more) sites. It is the process of transnational migration that provides the broad context for these relationships. In addition, transnational caregiving often takes place across considerable geographic distance.[2] What distinguishes the caregiving practised in this study from local (proximate) and translocal (across distance within the same nation state) varieties is that the families who participated in our research are involved in care exchange between members who live in countries distant and different from each other.

Transnational caregiving requires active kin relations. In this research we have focused, in particular, on the relationships between migrant children and their parents in the homeland. But it is important to note that, at any one time, these individuals may have numerous kin living in other countries, only some of whom they are in contact with. For care to be exchanged across distance and national borders, family members must not only be aware that overseas kin exist, they also need to have developed avenues of access to them.[3] Some links may be 'dormant' and could potentially be activated (or re-activated if they were once active but have since broken away) if and when the need arises.[4]

Against this backdrop of migration, national border crossings, geographic distance and active kin relations, a variety of patterns and types of transnational family caregiving emerge. The practices of transnational caregiving also take place over time and are played out within individual, family and migration life cycles or life courses. These care exchanges are mediated by a dialectic encompassing the capacity of individual members and their culturally informed sense of obligation to provide care, as well as the particularistic kin relationships and negotiated family commitments that people with specific family networks share. A diagrammatic representation of our model is presented in Figure 8.1.

This model illustrates the complex mix of motivations that inform the *exchange* of transnational caregiving practices that flow in both directions – from migrant to 'homeland' kin and vice versa. It is to the dialectic of capacity, obligation and negotiated commitments that we now turn.

Capacity (ability, opportunity)

Assuming the existence of active transnational kin relationships, one of the most influential factors affecting the motivation to participate in transnational caregiving is *capacity*. This dimension influences, and is influenced by, all other factors. *Capacity* refers to a myriad of issues that encompass an individual's opportunity and ability to engage in practices of transnational caregiving.

Capacity to participate in transnational caregiving can be impeded or facilitated by 'macro' structural factors, including migration policy and visa restrictions, employment policies, access to travel and telecommunication technologies, international relations between home and host

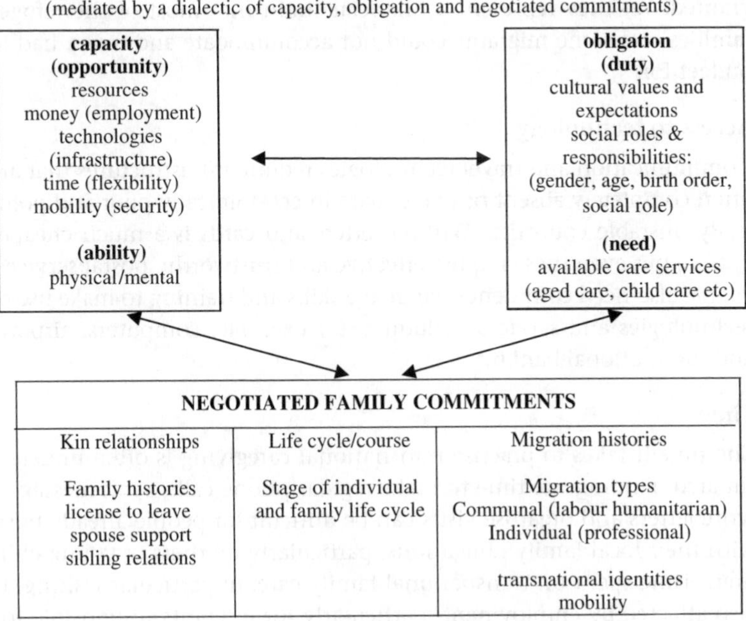

Figure 8.1. Transnational caregiving model

countries and the political stability and safety of relevant nations. Also playing a role in shaping capacity to care are the 'meso' or community factors, including the availability of local support and associations as well as welfare services and infrastructure. Finally, there are a number of 'micro' personal factors, in particular, available resources, including finances, employment status, language, health and time, but also willingness to allocate these resources to care exchanges and the perceptions of risk, safety and the effectiveness of caregivers. Willingness is in turn mediated by cultural constructions of obligation, need and negotiated commitments (discussed further below). Capacity, then, is primarily related to the availability and affordability of resources, including money, time, technology and mobility.

Finance

Participating in transnational caregiving requires money. Communication and travel technologies, essential to transnational caregiving, can be very costly. Acquiring and maintaining or accessing a telephone, computer, fax machine or mobile phone to send messages, as well as visiting and hosting visitors, are costs that many in our research took for granted. Yet they represent a financial strain that most of the refugee families and some migrants could not accommodate and many had to budget for.

Access to technology

Communication and travel technologies require infrastructures that are often completely absent or inadequate in economically poor and politically unstable countries. Writing letters and cards is a much cheaper option, but even these require effective and trustworthy postal services. People also need confidence, language skills and training to make use of technologies and services including, for example, computers, airports and international banking.

Time

The time it takes to practise transnational caregiving is often underestimated. Finding the time to make regular phone calls, send messages, write letters and organise visits can be difficult for people already busy with their local family obligations, particularly mothers of young children. Time spent on transnational family care, in particular visiting, is also affected by employment, particularly for migrants responsible for supporting a young family. Job flexibility is an important consideration as well, with many people, especially women, opting for part-time work to accommodate visits. Negotiating time off work and not being discriminated against in the workforce for taking leave are important policy issues.

Mobility

Clearly, having freedom of movement to visit kin across borders would appear to most people to qualify as a basic human right. Restrictions on one's mobility do not only come in the form of immigration and visa controls, however, and many people face inadequate provisions in both general employment and the airline industries to accommodate travel for the express purpose of providing care to kin. These examples highlight how transnational caregiving is a pertinent, though largely invisible, policy issue.

Ability

Considerable effort and skill is involved in practising transnational caregiving. Notwithstanding issues of access to resources and the specialised ability of using sophisticated technologies, a certain level of physical and mental health is required to provide most types of care. An individual's physical and mental ability is often affected by the ageing process. The heartache of migrants who can no longer converse with their dementing parents is palpable. As is the pain of a parent whose eyesight is no longer adequate to read messages or who has lost the agility needed to write them. Poor health, the psychological fear of flying, anxiety about negotiating airports (particularly for non-English speakers) as well as the sense of insecurity associated with 'global terrorism' are key reasons that inhibit people from undertaking travel to make visits, even if they can afford the time and money to do so.

The absence of capacity

When disability and lack of resources impede the practices of transnational care, individuals often choose to support other people who can more easily engage in caregiving (and thus engage in generalised reciprocity). For example, a parent might help to finance a local child to visit a distant sibling to help with the arrival of a new baby. Or a migrant might begin to phone and send messages to a sibling who has taken primary responsibility for caring for a physically or mentally ill parent. Allocating resources to transnational caregiving practices inevitably means spending money and time on caring for geographically distant kin that could be spent on local kin or some other activity. The tensions that sometimes arise over the resourcing of transnational caregiving are key factors in the negotiation of commitments, which are also often informed by cultural notions of obligation.

If the existence of capacity to care is a key factor in its delivery, does limited or absence of capacity to care affect an individual's motivation to provide care? This question is intimately connected to the notion of obligation, the availability of services and the history of negotiated commitments within families. We found that, in general, migration was not a valid excuse to completely avoid obligations to care. However, many individuals were not able to provide the extent of care, particularly personal care, that their parents hoped for. Some migrants, for example, were not able or willing to repatriate despite their parents' desire to have them living close by. We found, also, that the sense of obligation to care influenced capacity to care in powerful ways, for example, compelling people to find and spend the resources needed, as well as to manage any risks involved.

Obligation (cultural sense of duty/perceptions about need)

Sense of obligation has a significant impact on the practice of transnational caregiving, with people making decisions about resources, mobility and time in order to accommodate obligations to care. Obligation is the dimension that accounts for cultural understandings of transnational caregiving relations and highlights the central mediating effect of cultural values and expectations on kinship relations. Analysing practices that cross national borders inevitably leads to an awareness of cultural differences. The cultural sense of obligation influences all other dimensions – how, why and to what extent individuals care and who they care for, how they interpret mobility and distance, the way they manage their resources and so on.

Obligation is intimately tied to expectations and notions of appropriate social roles and responsibilities, as in, for example, being a 'dutiful' daughter or a 'good' father. This dimension refers to the cultural sense of moral duty and is implicated in notions of appropriate gender roles and birth order; the role of daughters as opposed to sons or of the first-born as opposed to the last-born. Here we might refer to cultural constructions of responsibility and expectation, but also how these might change as a result of migration and life experiences, particularly in a new country.

Obligation (in all its cultural manifestations) is also closely connected to levels of need and the relative needs of migrants and parents. Indeed, cultural understandings of caregiving both inform, and are informed by, the public and private organisation of care. For example, Blackman (2000) characterises certain European countries, including Denmark, Norway and England, as individual-oriented systems, in which most caregiving is highly organised and administered through government provided services. We would argue that the Netherlands and New Zealand approximate this type of system. In contrast, Blackman (2000) defines Italy, Ireland and Greece (and we would add Singapore, Iraq and Afghanistan) as more family-oriented systems, in which caregiving is largely provided within the family and community (see also Blackman, Brodhurst & Convery, 2001). Clearly, the level of available services impacts on the level of 'need' an individual has for care from kin, and this in turn informs the sense of 'obligation' to provide care. At the same time, the cultural notion of appropriate care impacts on the public development of service provision. It is important to note also that the perception of need is informed by local knowledge. This kind of awareness can be diminished through migration and may result in idealised or stereotyped notions about available care. Similarly, perceptions of need can be influenced by new experiences and practices in the host country, which may transform expectations and notions of obligation.

For many families, particularly those from Italy, Ireland, Iraq, Afghanistan and Singapore, there is a general expectation that elder care be provided within the family, such that public service provision of aged care in these countries is (more or less) limited. Family members from these places reported a very high sense of obligation to provide care, particularly personal (hands-on) care, to their kin. Many migrants openly described themselves as 'bad' children who failed to live up to their parent's expectations because, by migrating, they were unable to provide the level of personal care that was culturally expected. Similarly, parents felt they were somehow 'failing' their children by not being able to support them with childcare. Even moral and emotional care, for these families, is often closely linked to being proximate. These family members find themselves in the difficult predicament of negotiating, from a distance, the cultural expectations and obligations that dictate that appropriate care of kin requires regularly being with them.

Many migrants reported significant tensions surrounding decisions that involved moving parents from autonomous living arrangements, particularly living in their own homes, to improve their care support systems. Many parents, particularly the Italians and Irish, refused or tried to resist moving out of their homes, and several Italians also refused or resisted having home help, in the form of personal and nursing care provided by non-family members. Even the healthy and independent parents in these countries expressed some dismay at the idea of one day having to leave their homes. Their rejection of these forms of care indirectly emphasised the cultural expectation that they would be cared for by their children. Indeed, a number of Irish and Italian migrants returned to provide extended periods of personal care to dependent parents and to provide some respite to siblings who carried the primary responsibility for their parents. Similarly, many of the Iraqi and Afghan refugees wished they could sponsor their parents to migrate to Australia so that they could care for them.

Other migrants, especially the Dutch but including some Italians and Irish, resisted providing this level of personal and proximate care and assumed that their parents would move to assisted living arrangements. These individuals all cited their migration and local family commitments as reasons for not being able to provide their parent's preferred care options, although they all still actively exchanged what care they could through regular communication, visits and, in the case of Singaporeans, Iraqis and Afghans, financial support. These findings indicate that migration is used as a legitimate excuse regarding the degree and types of care, but it is rarely used as an excuse to cease care altogether. Obligations can also be ameliorated by the presence of local siblings if they take over the responsibility for the care of parents, or by the provision of services (home help or institutional accommodation). Nonetheless, even if parents are (objectively) well looked after, migrants from these families still often feel they are not fulfilling their duty as 'good' children and their parents also often view the situation as less than ideal.

Sometimes, obligation to care can impact on motivation to provide care in surprising ways. A migrant son decided to stop communicating and visiting his mother when she became severely demented and no longer recognised him. A migrant daughter made the difficult decision to stop visiting her mother as they both found the departures too painful to cope with. There were also many cases of migrants who cared for parents despite having very strained relations with them. Similarly, it was not uncommon for parents to care for migrants despite disapproving of their migration. In these cases, the sense of duty to discharge responsibilities to kin overshadowed any personal animosities, but did not necessarily make the experience any easier for the people involved.

While cultural background is a significant factor informing notions of 'preferred' and 'appropriate' aged care, it is important to note that many of the issues associated with intergenerational responsibilities and 'distant' care of the elderly are not easily differentiated according to national or ethnic group identity (Manderson, 1990). While we have highlighted some of the differences between the countries in our sample using Blackman's (2000) model, all the migrant-sending countries (the Netherlands and New Zealand, in particular, as well as contemporary Ireland, Italy and Singapore) also share a number of similarities that might categorise them as 'first-world' countries in the development sense. They are all relatively affluent, with well-educated populations, strong central governments and recent histories of governmental intervention in social economic affairs. As outlined in Chapter 3, these countries have also all recently been affected by similar demographic and social changes, including increasingly ageing and mobile populations, changing gender and work patterns and resultant family fragmentation and a tendency towards implementing neo-liberal reforms which have generally undermined state provision of the welfare sector (Harvey, 2005). Such similarities across the countries in our sample are as important to debates about aged care as are the diverse cultural characteristics between and within them.

In other words, it is reasonable to assume that there are commonalities in the experiences and needs of older parents across the globe. Further, the differences that exist are just as likely to result from gender, socio-economic, class and educational background, as from any factors relating to culture or country of residence. This said, our research shows that historical patterns in the common arrangement of care in the home-country, closely related to the level and provision of relevant services provided by the state, and reflected in culturally constructed notions of independence and preferred arrangements of care, have a significant impact on the experience of caring of the distant caregiver (Baldassar, Wilding & Baldock, 2006; Baldassar & Pesman, 2005).

Both capacity and obligation are key dimensions of the motivation to engage in transnational caregiving. But they are not the only factors, nor are they straightforward. Having the capacity and the cultural obligation to provide care does not necessarily, nor always, result in the provision or the acceptance of care. Capacity and obligation influence each other but are also mediated by a complex set of factors that we call negotiated family commitments.

Negotiated family commitments

Finch and Mason (1993) define negotiated family commitments as based in the history of relationships over time of the individual biographies involved in caregiving exchanges. We add to this dimension the migration histories and identities of parents, migrants and their kin. Thus, in our model of transnational caregiving, negotiated family commitments account for the particular kin relationships that develop over time, in the context of migration, and the fractured relations that often result. Migrants and their stay-behind kin invariably develop different perspectives on migration and this can cause tensions to develop between family members. Whether or not a migrant is given 'license to leave', to move away or to live at a distance, can be a major factor that influences the tenor of relationships (Baldassar, 2001, 2007). Similarly, the type of, and motivation for, migration influences family relationships and expectations in different ways. Further, people's views about migration (and the attendant license to leave) can change over time. A migration that was once accepted may become contentious, and vice versa, as a result of a crisis or simply changes in the life cycle (e.g., the Dutch parents becoming less willing to hide their emotions as they age). The notion of negotiated commitments allows us to consider the more personalised and intimate family dynamics that arise in transnational caregiving exchanges.

License to leave

Family relationships are influenced by the process of migration in many and profound ways. The manner in which a migrant's decision to migrate and settle in another country is received by her family informs their negotiated family commitments. How parents and siblings react to a family member's decision to migrate can have a profound effect on the way they feel about each other and the migration and, as a consequence, their expectations to give and receive care. In those cases where the migration has not been well accepted by family members, some degree of resentment towards the migrant is evident on the part of parents and local siblings, who feel the migrant, by departing, has abdicated his/her responsibilities for caregiving. Similarly, many migrants feel their decision to migrate forfeits their right to intervene in issues that concern their parents, and so they defer to local siblings on these matters. Migrants who do not have license to leave (from their families) often report feeling guilty (or being made to feel guilty), a situation that sometimes compels them to become vigilant and attentive transnational caregivers.

Of course, it is also conceivable that individuals might choose not to engage in transnational family caregiving exchanges. The nature of our methodology and sample meant that we did not encounter this situation in the families we examined, although it is not uncommon to hear of people who migrate in order 'to get away' from family and attendant obligations. Transnational caregiving is therefore highly influenced by how supportive significant kin are in facilitating its practice.

Supportive kin (spouse, siblings, children)

The level of support from individual family members, although influenced by capacity, need and cultural notions of obligation, is largely a consequence of family histories. The degree of support that spouses and children are prepared to provide can play an important role in enhancing or impeding an individual's motivation to exchange care transnationally. Local siblings may also have a significant impact on the migrant's capacity to exchange care, particularly as the composition and location of families change over time. For example, migrants usually stay in their parent's home during visits, but accommodation can become a fraught issue when parents no longer live in their own home. Similarly, when parents move in with a local child, or when disability limits a parent's ability to communicate across distance, local siblings may begin to control access to their parents. Divorce and re-partnering are also events that rupture and transform family relationships, having an impact on transnational care exchanges. Indeed, every dimension of transnational caregiving is dynamic and is affected by the stage in the migration process and the family life cycle.

Life cycle

Life cycle stage is a key consideration in negotiating family commitments as it can affect who takes on the primary role of caregiving responsibilities, when and how. For example, whether a transnational, translocal or local child, individuals who care for young and/or frail local kin may have less time to spare for transnational family members. As a result, the obligation to care may shift more heavily to a sibling who apparently has less (or less legitimate) demands on their time. Hagestad (1996) has argued that families should be seen as 'bundles of interwoven lives and interconnected timetables'. In this way, the stages in transnational family life cycles, mapped out through processes of migration across distance and borders, influence the type and pattern of transnational caregiving by informing the capacity, obligation and negotiated commitments on which caregiving is based.

The direction of care often mirrors processes of migration and the life course. For the recent, professional, more 'individually-oriented' migrants in our sample, caring flows primarily from parents to migrants in the early stages of migration when most migrants are establishing themselves. However, more care flows from migrants to parents when the latter age and become less independent. For the refugees and post-war, more 'communally-oriented' migrants, care, particularly in the form of remittances, usually flows from migrants to parents and may continue over time (in the case of refugees) or cease altogether (in the case of the post-war migrants). Thus, a focus on individual life course highlights the transitory and transformative nature of transnationalisms, and the way transnational migrant relations are intimately tied to family life cycles. To use Grillo's (2007) concept, transnationalisms are dynamic 'states of in-between-ness', highly processual in nature, that highlight migration as a fluid, transient process.

Transnational migration processes

Viewing transnational migration from the standpoint of the life course as a whole allows us to see how the exchange of care fluctuates throughout the life cycle, both within and across generations (Ackers & Dwyer, 2002). It also reveals that geographic mobility is often triggered by the need to give or receive care, rather than any 'rational economic', profit maximization motive commonly assumed in much migration literature (Ackers & Dwyer, 2002, p. 152). There is a growing recognition that all migrants (including refugees) move for mixed motivations, that it can be hard to disentangle political, social and economic reasons to move and that migrants are involved in a wide range of 'transnational' activities as migratory movements are not discrete, unilateral or linear. In charting people's care histories we often found multiple moves, in many directions, that did not always make the best economic sense.

This said, in terms of our findings, a useful distinction can be made between migrations that are motivated by family or communal concerns and those, generally more recent migrations, which are primarily motivated by individual, career and love interests. For an example of the communal migrations, we can point to some similarities between the Afghan and Iraqi refugees, all of whom migrated in the last decade, and the pre- and immediate post-war migrations of the Irish and Italians. While the former are political asylum seekers and the latter were mainly labour migrants, their processes of migration were through comparable chain or cluster networks. They migrated for the expressed purpose of alleviating the burden on family who remained behind through monetary support and the creation of the potential for family reunion migration. The economies of these sending areas relied heavily on migration and migrant remittances. Simply by departing, migrants reduced the competition for meagre jobs and resources at home. Importantly, these migrations are valued as appropriate decisions and generally these migrants have 'license to leave'. In their negotiated commitments with family, these migrants often find themselves called upon to support homeland kin (especially financially). The latter hold high expectations of receiving support and this forms a felt obligation for the migrant.[5]

Another characteristic of the more communal-motivated migrants is that they are members of ethnic communities or associations. Many foster relationships with kin and, more generally, co-nationals, in an attempt to increase the networks of care and support available to them as they negotiate the difficult processes of migration, as well as the tensions associated with settlement in a sometimes hostile social and political environment. Family and personal negotiated relationships take on a particular set of meanings in these historical and socio-cultural contexts. It is generally understood by all concerned that transnational communication and visits are not easily achievable. When one person from these communities makes contact with transnational kin or embarks on a visit, they, in part, communicate and visit on behalf of others.

Individual-oriented migrations, in contrast, involve homeland kin in different ways. Financial support often flows to the migrant as their families are generally more affluent. Some of these migrants, particularly those who leave to marry, are often not easily afforded license to leave by their families and this can place strain on kin relations. For others, (including, for example, the Dutch expats in our sample), migration is more likely to be an anticipated and accepted part of their career paths and life course. Moving to live in another country for these individuals is commonly seen as job or lifestyle mobility rather than traditional labour migration. We found that the way people constructed their own and other people's migration and identities coloured their experiences of transnational caregiving.

Transnational identity

As part of the study we asked participants about their ethnic identity and sense of belonging. As noted in Chapter 2, this information is important in informing our use of national categories ('Italian', 'Irish', 'Dutch', etc.) and their status in the minds of our informants. While not a focus of our research, we were able to glimpse some trends in the findings. Many people, particularly the professional migrants, do not identify as members of an immigrant or ethnic community, although the process of migration itself often results in a keen sense of their ethnic and national background and how these are perceived by the host society. They are also more likely to define themselves as 'global or world citizens' than as nationals, which has implications for how they perceive cultural obligations. Others, particularly the labour and humanitarian migrants, tend to identify more strongly with their town and region of ancestry rather than nation. They often describe themselves as belonging to an ethnic community in Australia.

We have some evidence to suggest that transnational caregiving practices can reinforce migrants' connections to both (geographically distant) family and home country, even when they are fully integrated within the host community. For parents and other homeland-based kin, these intersections across borders underpin continued transnational relationships and may result in attachments to the migrant's place of residence. We also found, however, that the absence of transnational identity (loosely defined as a sense of belonging to a particular place or places) does not appear to necessarily influence transnational caregiving practices.[6] In other words, not feeling a sense of attachment to a place does not impede caring for people in that place. Despite their varied transnational identities, the vast majority of our respondents expressed a high sense of obligation to provide care.

Transnational identity can be thought of as having (at least) three orientations: towards the homeland, towards the host setting or a mixture of both. So, for example, an individual might feel more Australian and less attached to their homeland or vice versa, or they may feel equally oriented to both places. Some migrants, particularly the recent professionals, reject any notion of traditional homeland culture as overbearing and backwards and, as discussed in Chapter 3, might be defined as 'cosmopolitans' (we return to this issue below). They tend to embrace life in Australia as liberating and emancipated. While this orientation ensures migrants will not repatriate despite parental disapproval, it does not stop migrants from participating in transnational caregiving.

Other migrants feel more affiliated with their place of origin and resist being Australian. These people often have great difficulty settling and long to be back in their homeland. Many parents fall into this category of rejecting Australia and feeling strongly that their homeland is superior. These are usually the parents who do not give their migrant children 'license to leave'. This scenario often makes parent visits particularly stressful. The question of the second generation is implicated in this dimension. Despite their parents' low identification with homeland, many grandchildren have strong ties to grandparents. Similarly, having a spouse who hates the homeland might not influence a person's sense of transnational identity but it might impact on their capacity to practise transnational caring (for example, spouses may withhold finances and/or resent time spent on transnational caregiving).

Many migrants feel an equal identification with Australia and their country of birth, and this is often the cause of some anguish about where they would like to be. If they have the capacity, it is these migrants who plan to spend if not equal, then extended amounts of time in the homeland. Transnational identity orientation might also change with the stages in migration and life cycle, being highly homeland-focused immediately following migration, particularly if migration was not a matter of choice. It may also grow over time to become strongest in later life.

Familiarity with mobility

A factor we found to be somewhat related to transnational identity, and one that appeared to have an impact on an individual's motivation to engage in transnational caregiving, was their familiarity with mobility and their identification as 'global citizens' or 'cosmopolitans'. People who identified in this way were more likely to be familiar with frequent travel, moving and living at a distance from kin. Consequently, they were potentially more aware of the need for support and the particular issues about how to provide it, as well as being more likely to cope with the stresses and strains of living apart from loved ones.

It is interesting to consider this view of mobility with reference to Jennifer Mason's (2004) notion of 'distant thinking' (a 'can-do' attitude to distance and kinship). Mason (1999, p. 170) examines different 'styles of reasoning about proximity, distance and kinship'. In her terms, a 'distant thinker' is someone who does not view distance as an impediment to functioning kin relationships and who has a view of distance as 'malleable' (p. 162). This is in contrast to 'local thinkers' who feel that proper or effective kin support is only possible locally (p. 170) and 'reluctant distant thinkers' who are willing to relate to kin over long distances only on a temporary basis (p. 167). We would argue that each of these 'types of thinking' is largely dependent on family histories, negotiated commitments, capacity and sense of obligation. Equally importantly, people might change their thinking depending on these factors over the course of the family life cycle.

The relationship (if any) between transnational identity and familiarity with mobility and the practice or motivation to engage in transnational caring is an interesting question and one that could provide a rich seam for future study. In the final section of this chapter, we provide a summary of our key findings and contributions to existing research.

Findings and contributions to research

Transnational caregiving is a largely invisible issue that needs to be acknowledged in theory and policy as well as at the level of lived experience. In a recent review article on sociology and care, Michael Fine (2005, p. 250) makes a compelling case that 'care needs to be made central to the sociological enterprise'. Care, he argues, 'is revealed as a central foundation of social life, a building block on which all subsequent social relations and processes depend' (Fine, 2005, p. 253). Fine is concerned that 'there is still little in the way of a sociological literature that sheds light on either pre-existent practices or the transformation of care and its social importance' (2005, p. 249). Despite going on to highlight key issues for a sociology of care, he does not mention caregiving in transnational contexts.[7] Similarly, while recognizing that 'forms of domestic life are changing dramatically, but [that] this has not led to a loss of commitment to care', Fiona Williams (2004) in her British study, *Rethinking Families*, does not consider the impact of migration on people's caregiving exchanges.[8] Her findings are in line with Barbara Pocock's (2003) important Australian study, *The Work/Life Collision*, which also overlooks transnational family relations.

While not cognizant of transnational caregiving, both William's and Pocock's comments on policy are pertinent. Williams (2004, p. 256) highlights the need for policy development relevant to new forms of domestic life, arguing that people are finding new caring arrangements to meet their emotional, social and economic requirements, but that 'policy needs to learn how to support them'. Pocock (2003, p. 13) asserts that governments affect the spheres of love, relationships and family; 'Governments, the media, employers, unions and community organizations powerfully affect the economic, social and cultural frameworks that naturalise, or makes strange, our ways of being at work, at home, and in our households', and, we would add, our ways of being transnationals. Work by Louise Ackers and her colleagues on transnational families in the European Union is one of few to consider policy implications (Ackers, 1998; Ackers & Dwyer, 2002; Ackers & Stalford, 2004). A key aim of our book is to raise awareness about caregiving as it is practised in transnational contexts and the relevant political, both public and private, issues. Families caring across distance need to be supported through appropriate social, community, employment, legal, migration and health policies; at present all of these sectors remain, for the most part, entirely unaware of transnational care practices and problems.

Even analyses of transnationalism have tended not to examine the nexus between the public arena and the domestic sphere, with most studies focusing on the former, and particularly the political or citizenship (primarily male) sphere at the expense of the latter, including household, family and the lives of ordinary women and men (see Gardner and Grillo, 2002). With its focus on family caregiving practices, our work is an example of 'transnationalism from below'. We feature the everyday realities, concerns and experiences of transnationals and how they are affected by governments and state policies in their family caregiving. For example, there has been some reference especially in the media about the potential threat of split political allegiances presented by dual or multiple citizenships. Our research found that people are primarily interested in multiple citizenships to facilitate family caregiving relations and are not necessarily concerned with national allegiance.[9]

The notion that globalisation is providing less impediments to transnational networks and the related question of the demise of the nation-state did not appear especially evident in our study. Rather, individuals had to contend with real barriers to caregiving presented by borders and distance. This finding resonates with Torpey's (2000, p. 13) criticism of Anderson's concept of 'imagined communities', which he argues tends to ignore 'the extent to which identities must become codified and institutionalised in order to become socially significant'. Torpey defines Anderson's view 'subjectivist' for not taking into account the way identities are 'anchored in law and policy' (through, for example, passports). Both home and host states shape the activities of transnationals through obstacles of immigration law, employment and welfare policies and services (see Chapter 7).

Borders remain an obstacle, affecting the practice of transnational caregiving. Our work highlights the impact that borders have on transnational practices of care and therefore provides an argument for the distinction between translocal and transnational family relations. Caring across national borders has consequences that are absent in families who are simply caring across distances, however large, but within the borders of a single nation. At the same time, we distinguish between types of transnationals. The majority of our sample are professional, affluent, 'individually-oriented' migrants and their families who approximate the category of transnationals defined in the literature as 'cosmopolitans' (see Hannerz, 1992; Werbner, 1999; Skrbis, Kendall & Woodward, 2004). The remainder of the sample, while still engaged in transnational relations, are working class and more 'communally-oriented'.

As noted in Chapter 3, Werbner (1997, p. 12) identifies a similar class-based distinction, arguing that the differences between the two groups refer to the types and levels of transnational engagement;[10] '[the proletarian variety] are those whose loyalties are anchored in translocal social networks' as compared with cosmopolitans whose loyalties are located in 'the global ecumene'. We found, however, that even the more cosmopolitan participants in our samples, many of whom are engaged in the global ecumene (particularly the expats), are still also focused on quotidian transnational family networks. They approximate Lamont and Aksartova's (2002) 'everyday, practical cosmopolitans'. Further, while many of the migrants in our sample display 'cosmopolitan' attributes, including what Beck (2002, p. 30) calls 'a new kind of identity and politics as well as a new kind of everyday space-time experience and of human sociability', they also often belong to families whose members are not equally globally connected, with parents, for example, who are far less mobile and less versed in 'a variety of cultural repertoires' (Hall, 2002,p. 26). In other words, even the most cosmopolitan of migrants is concerned with ordinary, everyday activities of people whose relationships stretch across distance. All transnationals engaged in family caring relations have to negotiate family commitments, obligations and capacity.

Similarly, migration and frequent travel are not always, nor only, the result of economic considerations. An analysis of long-distance care demands an approach that examines the family life cycle and caring, along with economics, as components of migration decisions. Migratory moves may be prompted (or cancelled) by the need to give or receive care and people undertake visits in order to provide care to loved ones (Ackers & Dwyer, 2002, p. 167). These decisions are not so much motivated by weighing costs and benefits to maximize economic gain, as by what Pocock (2003, p. 16) calls 'caring decisions which rely upon altruism and the sacrifice of personal interest to the interests of another'. The mobility of care is thus located in the realm of the economies of kinship, including but not limited to, the exchange of economic resources within families. Further, some migrants undertake lengthy repatriations due to nostalgia and homesickness as a way of caring for their own health and wellbeing.

Transnationals include migrants, their partners and children as well as non-migrants and their extended kin. Our contribution to transmigration studies is to emphasise that migration is a process that continues beyond the migrants' immediate settlement stage and has an ongoing impact on their kin and associates in both their old and new homelands. Existing definitions of, and research on, transnationalism tend to privilege 'the process by which *immigrants* forge and sustain multi-stranded social relations that link together their societies of origin and settlement' (to use a frequently quoted definition by Basch, Glick Schiller & Szanton Blanc, 1994, p. 7, emphasis added). As a result, various other actors in the networks created by transnational migration are rendered somewhat invisible. Our study highlights the fact that migrants who have actually moved across borders are not the only transnationals. So too are the migrant's kith and kin who stayed behind and their partners and children who are not themselves migrants.

Thinking about transnationalism in this way *broadens* the definition. The difficulty of finding an appropriate term to describe these people – examples include 'stay behinds', 'homeland kin', 'home-based', 'non-migrants', 'second-generation' – is indicative of a gap in scholarship. The point here is that the focus of diaspora and transnationalism studies tends to be more heavily weighted towards host country settings. The sending areas become relatively peripheral, excepting a concern with development issues and remittances (Ostergaard-Nielsen, 2003). An examination of transnational care practices spreads the focus more evenly between the two (or more) sites, through an examination of the ways family members in both (or more) places are oriented towards, and relate to, each other. Further, transnational caregiving relationships conducted by families caring across borders involve not only individuals but also collectivities (local communities) in both the sending and receiving areas as well as relevant public institutions. This focus is an avenue to explore the largely unexamined impact that sending areas have on transnationalisms.

The elderly are not a homogenous group. Despite contemporary migration policies viewing elderly migrants as a potential burden to their families, communities and the state, many researchers have found that elderly parents are active caregivers (see for example, Arber & Evandrou, 1993; Hareven & Adams, 1982; Ackers, 1998; Ackers & Dwyer, 2002; Ackers & Stalford, 2004). Katz (1996, pp. 3–4) argues that a 'Western approach to professional gerontology problematises the aged individual and the old body rather than the social conditions of ageing', pointing out that 'there is no universal old person' and that the 'Aging process is heterogenous and indeterminate'. Our study found that the emotional and financial support provided by parents, in particular in the early settlement stage of migration, was significant in both quantity and impact. Parents continue such support so long as they are financially able and in reasonable physical and mental health.

In spite of the vast distances and national borders that separate them after migration, our research shows that transnational family members do continue to exchange care.

As with local and translocal family caregiving, a diversity of types of caregiving are exchanged between family members. Moral and emotional support is often continuously and constantly exchanged in transnational kin relationships as people care for, and about, each other. A variety of levels of financial and practical assistance is also exchanged. And even accommodation and personal (hands on) care are provided by transnational kin during visits. Moments of co-presence during visits offer important opportunities for building and maintaining family relationships across borders, distance and time.

The practice of transnational caregiving is informed by a dialectic of capacity, obligation and negotiated commitments. The broad pattern of transnational caregiving can be described as involving two key types of care practice: *routine*, day-to-day caring, characterised by regular contact; and *crisis*, key event caring, involving an increase in time, effort and resources. The frequency, mode and type of care exchanged in both routine and crisis transnational caregiving is dependent on an individual carer's capacity, sense of cultural obligation and negotiated family commitments. For some, particularly the recent, younger, more affluent and mobile professional families, routine caregiving occurs on a daily basis through phone calls, SMS and email messages containing the details of daily life that ensures family members feel connected, in touch and informed about each others' health and wellbeing. Members in these families are also more likely to engage in routine visits every year or so. For others, particularly the families of the humanitarian and labour migrants, routine caregiving is commonly characterised by weekly or monthly phone calls and occasional letters and postcards. For these family members, visits are more likely to be infrequent special events.

In contrast to the regular and continuous exchanges of routine caregiving, crisis transnational caregiving involves additional time, effort and resources. We found that across all samples (thus cross-culturally) there are three key stages in the transnational family life cycle that invoke increased levels of caregiving – the period immediately following migration, time around the birth of babies and when parents lose their independence and become frail or ill. Only the first stage is unique to transnational (and translocal) caregiving, (although the other stages, while not unique to migration contexts, are arguably more difficult to negotiate transnationally). We also found that the ability to fulfil the need for greater contact is subject to intra-sample variations such as the limitations of migration category, life cycle stage and access to financial resources. The need to provide care is affected by cross-cultural variations in the construction of the family, age-set roles and gendered obligations to care, as well as by availability of other caregiving relatives in the home and host countries.

New communication technologies assist in the practice of transnational caregiving; they also change the pattern and frequency of communication across distance. The practices of transnational caregiving are largely dependent on, and greatly influenced by, communication technologies, arguably to a greater degree than in local and translocal families. The transformations in these technologies have resulted in major changes in methods and patterns of long-distance communication over time. While letter writing and postal services were the main vehicles of moral and emotional, as well as practical and financial, support up until the 1980s, they have now been significantly replaced by telephone calls and email. Postal services now tend to be reserved for 'special' situations when people want to show extra 'care', for example, in relaying some particularly distressing news by letter. In addition, many people send greeting cards, faxes, photos, newspaper clippings and children's drawings, in an effort to offer a more corporeal form of contact through the touch, sight and smell of the paper and its contents. Gift exchange is also an important part of transnational caregiving and great pleasure usually accompanies the arrival of a parcel from overseas kin.

Regular telephone calls, emails and SMS texting now characterise transnational caregiving. Along with emotional and moral support, these appear to be excellent avenues for the provision of practical support, decisions can be discussed and advice sought and given in real time or close to it, without the delays of the post and, most importantly, at the time it is most needed. Extended networks of kin are more likely to be involved in transnational communication through the use of the newer technologies than was common when letter writing was the main vehicle of exchange. While it was usually the parents (mainly, but not always, the mothers) who were the main letter writers, forming a 'hub' or centre for the receipt and dispersal of news from and to the extended family, today transnational caregiving commonly involves greater numbers of people in a 'dispersed network' of communication.

We also found that new technologies and their greater availability have increased the sense of obligation and expectation to participate in transnational caregiving exchange. Further, by increasing the sense of 'presence' of family across time and space, communication technologies also appear to increase the desire to visit. The families in our study make use of all the forms of 'staying in touch' across distance that are available to them. Indeed, as noted in Chapter 6, a key finding of our study is the emergence of a new set of factors that transnational families must take account of in their negotiation of commitments to care across borders. In addition to past histories of kin relations, it appears that relations to the material world – of technologies and resources – are just as important in determining perceptions of the obligation to care. That is, they impact upon the capacity to care, which in turn informs the sense of obligation to care.

Future research questions

We hope that our research has made a useful contribution to reconceptualising family studies, migration studies and ageing studies by thinking about the ways in which these various fields intersect in people's lives. Nevertheless, there are many questions that remain unanswered, and which require further research.

The implications of gender for transnational caregiving have been dealt with to some extent in this study. For example, we have pointed to the ways in which gendered expectations vary across different national groups and the potential struggles for power within family networks that are often linked to gendered patterns of caregiving and resource management. However, we have only just begun to touch on the implications of gender for transnational caregiving and transnational family life. In part, this is because of our specific focus on relationships between parents and their adult migrant children. Our study so far suggests that other relationships, such as sibling and spousal relationships, are also significant and deserve further attention. We would like to see these issues addressed more systematically. For example, are there gendered differences in the willingness to take responsibility for routine, as opposed to crisis, transnational care? Or in parental, as opposed to extended-kin caregiving? Do men and women experience cultural obligations to care in different ways, and with different consequences for their sense of being 'good kin'? What are the consequences of transnational family obligations for intimate relations, child-care and work-life balance within the nuclear family household? Also, does the increased role of technology in family relations conducted transnationally have an impact on the preparedness and capacity of either men or women to care and conduct kinwork?

The implications of technology for transnational family life and caregiving comprise another important area for future research. We are as yet unable to answer the question of whether co-presence is essential to the maintenance of relationships, or whether information and communication technologies, by enabling families to share time, to some extent replace the need to share space. The use of mobile phones, e-mail and digital cameras has brought an immediacy to transnational relations that was inconceivable to migrants in the past. What are the consequences of this transformation for the maintenance of family relations – both transnationally and locally? Has the opportunity to 'keep in touch' resulted in a much stronger pressure or moral obligation to keep in touch more often, or in more intense ways? Do the new opportunities to communicate impose an obligation to connect with a larger and more diverse network of extended kin? If so, what are the implications of these obligations for local family and community life? Are transnational families being required to spend more of their limited time and resources on transnational networks, to the detriment of local communities and social life? Or do transnational networks operate to 'add value' to local family and community life? Also, further an in-depth ethnographic research would be useful to examine how transnational families are 'imagined' across space and time by their various members – who, like fellow citizens, are not in regular face-to-face contact but are nevertheless bound by a sense of belonging to a transcending, perhaps virtual, entity, in this case not the nation, but the transnational family.

We should also not assume that all migrants participate in transnational family life, or that all transnational families are the same. In terms of different types of migrants, there are many questions that we would like to see explored more deeply. For example, are transnational family networks more common among relatively wealthy, professional migrants, or do refugee and proletarian migrants also have access to transnational family life? Does access to transnational kin networks improve the settlement experiences of migrants? Or do they impede the capacity and willingness to engage with local communities and networks? In this sense, it would be useful to see a comparison of other groups of migrants, including a more specific focus on expatriates and on labour migrants, as well as people who repatriate to their home country, and people who choose to migrate after retirement in order to be closer to family or in order to separate away from family obligations. An examination of migrants who relocate specifically as an economic strategy, sending remittances back home, would also be an important addition to our study.

In addition to the above suggestions for future research, we hope that the findings of our research will contribute to practical initiatives that might aid the lives of members of transnational families. This is clearly an aspect of everyday life that is currently ignored by policymakers, in spite of having an effect on vast numbers of people. Yet adjustments to employment and migration policies could have significant positive impacts on the lives of transnational kin. Also, it would be encouraging to see more transnational resources available for people to learn about a range of effective strategies for maintaining long-distance relationships. The participants in our study often commented that they had not thought about these issues before. We feel certain that they and others like them would benefit from a forum where they could share their experiences and exchange information regarding strategies for maintaining family and care relationships more effectively across distance. Further research, services and initiatives that assist transnational families to access communication, information and advice could be of enormous benefit to many who are currently struggling to solve the problems of transnational caregiving alone.

Notes

1 Introduction: Transnational Caregiving

1. See, for example, Opie (1992), Finch and Mason (1993), Bengtson, Rosenthal and Burton (1995) and Batrouney and Stone (1998).
2. See, for example, Finch and Groves (1983), Ungerson (1990), Opie (1994) and Stein et al. (1998).
3. This is also acknowledged by Finch and Mason (1993, p. 163).
4. For discussions of filial piety see, for example, Treas and Wang (1993) and Bengtson et al. (2000), especially Chapter 3.
5. The closest capital city to Perth, Adelaide in South Australia, is about a 3 hour flight away.
6. See, for example, Bhabha (1989), Appadurai (1991), Bottomley (1992), Ang (1994), Basch, Glick Schiller and Szanton Blanc (1994), Gilroy (1994) and Verdery (1994).
7. Notable exceptions include the work of Louise Ackers (Ackers, 1998; Ackers and Dwyer, 2002; Ackers and Stalford, 2004), Karen Fog Olwig (2002, 2003) and the volume by Bryceson and Vuorela (2002).
8. As Mahler (1998, p. 82) has noted in a critical discussion of research on transnationalism, it may 'yield detailed information on a limited set of activities and practices, not a clear picture of player's participation in all the activities people engage in'.
9. For a specific reference to cumulative commitments, see Finch (1989, pp. 201–205).

2 Researching Transnational Caregiving

1. For an extended description of Italian examples see Baldassar (2001).
2. Such Bilateral Agreements were drawn up between Australia and various home countries; this involved a sharing of cost of transportation and initial settlement.
3. If people have married abroad, foreign spouses enter under family reunion spouse visas; otherwise the Australian party can sponsor his or her intended partner under a prospective spouse visa. Australia is one of a very few countries to recognise *de facto* relationships, and also has special entry categories for same sex couples. Central to immigration policy in this area is the necessity of proving that the relationship is genuine, and does not involve a 'sham' marriage (see Crock, 1998, Chapter 5).
4. The Department of Immigration and Multicultural and Indigenous Affairs (DIMIA) usually sets the benchmark score.
5. Current shortages of skilled workers in Australia have led to increased efforts by the Australian federal government to bring in skilled migrants, with a target set for 2006 of 69.6 per cent of the non-humanitarian intake. In the

event this target is reached (which is unlikely), skilled migration would become a larger component of immigration than family reunion. To attract more skilled migrants, the MODL list and the Working Holidaymaker Scheme (for temporary tourist workers) are reported to be undergoing expansion (see, for example, *The Weekend Australian*, 5–6 March 2005; *The West Australian*, 15 April 2005, p. 4).

6. A total of seven parent interviews (one with Irish, two with Dutch, four with Italian parents) were conducted while the parents were visiting in Australia. This led to invitations to visit these parents again in their home country.
7. We also encountered instances of parents dying before their scheduled interview took place.
8. Several female migrants in the Dutch sample, whose parents were divorced, preferred us to contact their mothers, because of a sense of loyalty to their mothers and because their contact with their fathers was less extensive. This finding is comparable to Finch and Mason's (1990) regarding family relations after divorce.

3 Contexts of Migration and Aged Care: Research Case Studies

1. The section of this chapter entitled 'Singapore migrants and their parents' was written with Anita Quigley. The section entitled 'Refugees from Iraq and Afghanistan' was written with Zahra Kamalkhani.
2. Information about this group was drawn from Baldassar's previous research conducted during 1987–1989, 1993 and 1999–2000 with families in Sicily and the Veneto region in Italy and their kin in Perth and Queensland (Baldassar, 2001). The project included extensive participant observation and over 40 in-depth interviews. While not a formal part of our current project on transnational caregiving, data concerning four Italian families have been included for two main reasons: Baldassar's previous research, like Baldock's (1999, 2000), informed the development of our ideas about long-distance care; also, the transnational caring experiences of this cohort provide a useful comparison with the mainly recent migrants (post-1970s) that form the bulk of the Italian sample.
3. Interviewee 32122.
4. Interviewee 32161.
5. Interviewee 42061.
6. Interviewee 30032.
7. In 1947, only 2174 Australian residents had been born in the Netherlands; immigration to Australia dropped again after 1961, with new arrivals now between 250 and 500 annually (Jupp, 2001, p. 260). Even career migration within the Netherlands itself is a relatively infrequent phenomenon (Smits, 2001, p. 555), let alone emigration abroad.
8. Interviewee 31122.
9. Interviewee 31101.
10. There were no widowed parents in this group, but in two cases parents were divorced and only the mothers visited their migrant children.

11. Until very recently, the Dutch government required that Dutch nationals, who sought to take out citizenship of another country for reasons of marriage, renounced their Dutch citizenship. New legislation was introduced in 2003 that allows people in this category to reapply for Dutch citizenship, thereby effectively opening the possibility of dual citizenship.
12. Interviewee 41151.
13. There was one widowed father in this group and in the case of one expatriate couple both had divorced parents and only the mothers visited.
14. Interviewee 41032.
15. A group of Dutch parents has set up a fortnightly playgroup for their children, where parents meet and children play. This group includes expatriates and some migrants, all keen to ensure that their children have opportunities to speak Dutch.
16. *Mantelzorg*: Informal domiciliary care by adult children, other relatives or neighbours.
17. As noted by Timmermans (1996, p. 17) the percentage of older people living in intergenerational households declined between 1976 and 1993 from 16 per cent to 9 per cent; they are now mainly found in rural areas away from the major cities.
18. Interviewee 35071.
19. This usually means those who migrated prior to 2001, which includes the entire New Zealand sample.
20. At the time of the 2001 Census, WA had 10,270 Singapore-born migrants, NSW had 8510, Victoria 7610 and Queensland 4520 (see also Quigley & Menon, 2004).
21. Until the 1970s, the bulk of international students came under the Colombo Plan and other foreign aid programs, from 1974–1990 they enjoyed the same fee-free status as their Australian counterparts, but since 1990 all international students (more than 50,000 in 2001) pay full fees (Quigley & Menon, 2004, p. 255).
22. Interviewee 43022.
23. Interviewee 33111.
24. See, for example, Sullivan & Gunasekaran (1994); Quigley & Menon (2004).
25. Interviewee 43022.
26. Interviewee 33121.
27. Interviewee 43051.
28. Interviewee 43022.
29. Interviewee 33072.
30. This brief summary of Iraqi migration history is also documented in Kamalkhani (2004, p. 239).
31. See http://www.immi.gov.au [accessed June 2004].
32. Zahra Kamalkhani continued her research of these refugees in a project among the third wave of Afghan refugees, namely Hazara people who lived as TPVs in Western Australia. In this context she conducted a further 15 extended interviews with Hazara men in Albany and five of their family members in Iran.
33. For an overview of conditions of Iraqi refugees in Iran, see the entry on 'Iran' at: http://www.refugees.org [accessed September 2005].

34. Those who overstay their visa are then considered illegal residents and pay a large penalty before they are able to receive exit visas to join their relatives in Australia.
35. Landlords expect payment of a large bond, which they are often reluctant to pay back when refugees want to leave and join their relatives in Australia.

4 Transnational Caregiving between the Generations

1. Interviewee 34091.
2. In two instances, parents had gained refugee status in Australia, but some of their adult children were still living in transit countries. One of these also received transnational financial support.
3. Interviewee 34141.
4. See Chapter 7 for information about the costs of family reunion.
5. Interviewee 34032.
6. Interviewee 34071.
7. One Iraqi who had arranged to sponsor a fiancé from Iran said this had cost him $50,000 (Aust.).
8. Interviewee 33062.
9. Interviewee 43012.
10. Interviewee 43051.
11. Interestingly, she sent this money to ensure her mother could afford a few luxuries for herself; however, her mother usually spent it on presents for her local grandchildren.
12. Interviewee 32092.
13. Interviewee 31012.
14. Interviewee 40092.
15. Interviewee 30092.
16. Interviewee 35101.
17. Interviewee 40822.
18. Interviewee 30122.
19. Interviewee 32102.
20. Interviewee 45061.
21. Interviewee 31041.
22. Interviewee 30112.
23. Interviewee 32032.
24. One woman took the extraordinary step of hiding her mother's passport while her mother was visiting her in Perth so that she could not leave.
25. Interviewee 30061.
26. Interviewee 32092.
27. Interviewee 31012.
28. Interviewee 31112.
29. Interviewee 31112.
30. Interviewee 41092.
31. Interviewee 41122.
32. Interviewee 41102.
33. Interviewee 31042.
34. Interviewee 34081.
34. Interviewee 31041.

35. Interviewee 34062.
36. Interviewee 34032.
37. Interviewee 34091.
38. Interviewee 45011.
39. Interviewee 30081.
40. Interviewee 41141.
41. Interviewee 31112.
42. Interviewee 33022.
43. Interviewee 30092.
44. The latter was especially common among Singaporean migrants, who were otherwise not inclined to provide much practical support during visits.
45. Interviewee 35021.
46. Interviewee 34141.
47. Interviewee 43041.
48. Interviewee 33032.
49. Interviewee 33042.
50. Interviewee 43022.
51. When Singaporean parents did not have sons, daughters were expected to take responsibility.
52. Interviewee 32041.
53. Interviewee 40032.
54. Interviewee 33082.
55. Interviewee 43041.
56. Interviewee 43051.
57. There was only one exception to this: one respondent, who came from a Nigri-Sembila background, noted that in her family the eldest daughter was the most important as caregiver.
58. Interviewee 33051.
59. Interviewee 43051.
60. Interviewee 33102.
61. Interviewee 33082.
62. Interviewee 34052.
63. Interviewee 34032.
64. Interviewee 34012.
65. Interviewee 34132.
66. Her father died one year after she completed her studies, and she said 'so literally for ten years I was working and looking after my mum and my family'.
67. Interviewee 33082.
68. Interviewee 31072.
69. It is worth noting that many parents of long-distance migrant daughters gave them licence to leave because they migrated for love, in other words, they did what women are supposed to do, 'follow a man'. However, if and when their daughter's relationship broke up, such parents often exercised subtle pressure on these daughters to return home.
70. Interviewee 30061.
71. Interviewee 45072.
72. Interviewee 30012.
73. Interviewee 31012.
74. Interviewee 43051.

5 Communicating Across Borders

1. The lack of attention to differently abled bodies in the design of communication technologies has rarely been discussed, but see Goggin & Newell (2003, 2006) and Keating & Mirus (2003).
2. Interviewee 31021.
3. Interviewee 31072.
4. Interviewee 41062.
5. Interviewee 40052.
6. Interviewee 32012.
7. Interviewee 30192.
8. Interviewee 30112.
9. Interviewee 32102.
10. Interviewee 31052.
11. Interviewee 31051.
12. Interviewee 32022.
13. Interviewee 33082.
14. Interviewee 32092.
15. Interviewee 32131.
16. Interviewee 30092.
17. Interviewee 31101.
18. Interviewee 35092.
19. Interviewee 41051.
20. Interviewee 35092.
21. Interviewee 35092.
22. Interviewee 31101.
23. Interviewee 32102.
24. Interviewee 41032.
25. Interviewee 32112.
26. Interviewee 40052.
27. Interviewee 30092.
28. Interviewee 30021.
29. Interviewee 31101.
30. Interviewee 31112.
31. Interviewee 41092.
32. Interviewee 42022.
33. Interviewee 30112.
34. Interviewee 31051.
35. Interviewee 31202.
36. For more detailed discussions of the gendered uses and control of telephones and other household communication technologies, see, for example, Pool (1977), Kramarae (1988), Betteridge (1997), Smoreda & Licoppe (2000), Lacohee & Anderson (2001) and Na (2001).
37. Interviewee 31012.
38. Interviewee 43022.
39. Interviewee 31051.
40. Interviewee 31161.
41. Interviewee 34122.
42. Interviewee 34122.

43. Interviewee 34141.
44. Interviewee 34091.
45. Interviewee 34122.
46. Interviewee 34121.
47. Interviewee 34122.
48. Interviewee 44062.
49. Interviewee 34072.
50. Interviewee 44122.
51. Interviewee 34141.
52. Interviewee 34102.
53. Interviewee 30052.

6 The Role of Visits

1. Italian provincial transnational migrant associations organise special package tours to countries of emigration which enable homeland kin to visit their migrant relatives. For example, the Italian provincial *nel Mondo* associations organise annual trips around Australia. Each trip is planned to coincide with a reunion festival held in localities where large numbers of migrants have settled. The organised tours guarantee ample English language support and the comfort provided by co-national travelling companions for the mainly Italian-speaking visitors.
2. There is a considerable literature on tourist visits that locate tourism within transnational social fields, see, for example, Cohen (1972), Urry (1990), Williams (1995) and Duval (2003, 2004).
3. Interviewee 33072.
4. At the time of writing (2006) an off-peak Perth–Singapore return flight was about $900, similar to an off-peak Perth–New Zealand flight at around $800, compared to approximately $2000 for return flights between Perth and European capitals.
5. For a discussion of 'Asian values' see Yang (1988), Barr (2000, 2002) and the *Shared Values, White Paper* (1991), Singapore National Printers, Singapore.
6. For a discussion of 'Italian family values' in the Australian migration context, see Bertelli (1985), Rowland (1991), Castles et al. (1992), MacKinnon & Nelli (1996, p. 74), Contessa (2000), Chiro & Smolicz (2002) and Baldassar, Wilding & Baldock (2006).
7. In addition, many of the festivals may not be celebrated in Australia, which is a predominantly Christian country, unlike Singapore which has national holidays for major festivals in various religious traditions including Christian, Muslim and Buddhist.
8. Interviewee 32161.
9. Interviewee 33051.
10. Interviewee 32042.
11. Interviewee 31072.
12. Interviewee 35061.
13. Interviewee 34012.
14. Interviewee 32102.
15. Interviewee 30061.

16. Interviewee 31152.
17. Interviewee 31012.
18. Interviewee 43051.
19. Interviewee 30112.
20. The distinction between consociate and contemporary draws on Schutz's (1967) use of these terms – consociates share co-presence, contemporaries share time; the people you know are your consociates, the people you only know about are your contemporaries.
21. For a definition and discussion of diaspora see Cohen (1997), and for Italian diaspora, see Gabaccia (2000).
22. Interestingly, nostalgia is derived from the Greek verb *nostos* – to return.
23. This mirrors Mitchell's (1959) thesis that financial considerations (the necessary conditions) alone are not enough to motivate a migration; there are always additional, non-financial reasons (the 'sufficient conditions', such as weather, lifestyle and family) that ultimately tip the balance in favour of a move.
24. Interviewee 31062.
25. Interviewee 32112.
26. Interviewee 34072.
27. Interviewee 34032.
28. Interviewee 34061.
29. Interviewee 34132.
30. Interviewee 31242.
31. Interviewee 41192.
32. Interviewee 31012.
33. Interviewee 34041.
34. Interviewee 32152.
35. Interviewee 31201.
36. Interviewee 31042.
37. Interviewee 32102.
38. Interviewee 31152.

7 Implications for Policy

1. See, for example, http://www.workplace.gov.au/workplace/Category/SchemesInitiatives/WorkFamily/Carersleave.htm [accessed June 2006]; http://www.actu.asn.au/papers/history_parental_leave.html [accessed June 2006].
2. The opposite was also the case; we came across one Italian migrant family repatriating to care for an ageing parent because they had been told they would be disinherited unless they did so.
3. An important example of a religious institutional setting, which helps to maintain a community language, is the Orthodox Greek Church's custom of conducting its services in the Greek language. The use of Greek in this 'public' institution has been identified as an important factor in the retention of Greek language use in Australia (see Clyne, 1972; Clyne & Kipp, 2002). In contrast, the optional use of Italian in the Catholic Church in Australia, where masses are primarily conducted in English, has been linked

to Italian language decline (see Smolicz, 1983; Bettoni & Rubino, 1996; Rubino, 2000; Baldassar & Pesman, 2005).

4. We draw here on papers presented at a Symposium entitled 'Maintaining minority languages in a transnational world: Australian and European perspectives', held 5–6 June 2003, University of Western Australia. See also Pauwels (2006).
5. Of course, only in the United Kingdom where – like in Australia – English is the taken-for-granted majority language, would such a situation not occur.
6. Examples are: *Family Caregiver Alliance*, USA (www.caregiver.org, accessed June 2006); *National Alliance for Caregiving*, USA (www.caregiving.org, accessed June 2006); *Carers UK* (carersonline.org.uk, accessed June 2006); *The Carers Association*, Republic of Ireland (www.carersireland.com, accessed June, 2006); *Alles over Mantelzorg*, Netherlands (www.demantelzorger.nl, accessed June 2006). We thank Marja Pijl (2004) for this information. See also Spencer-Cingöz (1998).
7. See www.caregiver.org [accessed June 2006].
8. See http://www.aarp.org/families/caregiving/caring_parents/a2003-10-27-caregiving-longdistance.html [accessed June 2006].
9. See www.caremanager.org/displayassociationlinks.cfm, a website with nearly 300,000 links to relevant entries [accessed June 2006].
10. This information was given to the authors in personal communication.
11. Pijl (2004) suggests that transnational carers should collect names and telephone numbers of doctors and other relevant professional caregivers when they visit back home, so they can remain in contact with these experts from afar.
12. See www.dimia.gov.au [accessed June 2006].
13. There are also a multitude of rules as to where migrants and refugees must be located (onshore or offshore) at the time their permanent visas become available (Burn & Reich, 2005), which limit mobility during the first few years of residency.
14. Up-to-date information regarding current Reciprocal Health Care Agreement is available at http://www.medicareaustralia.gov.au/yourhealth/going_overseas/index.htm [accessed June 2006].
15. This is stipulated in the award conditions of the *Australian Workplace Relations Act 1996*, but it is possible for employers and employees to negotiate alternatives in Workplace Agreements, including additional periods of unpaid leave.
16. One migrant mentioned the existence of an insurance scheme in the 1970s which provided finance for air travel in the case of emergencies. The premiums were quite high and eventually the scheme folded.
17. Only a person who is included in another person's passport does not need to apply separately. In that case there is only one single fee.
18. Interviewee 33012.
19. Interviewee 33042.
20. Interviewee 43022.
21. Substantial changes were introduced in the category of parent visas in 1998, greatly restricting eligibility for parents of working age, and increasing health charges and assurance of support bond (Burn & Reich, 2005, p. 227).

22. For example, in 1997 nearly a quarter of British pensioners living overseas received their state pension in Australia (King, Warnes & Williams, 2000, p. 25). It should be noted though that the percentage of British pensioners in Australia has decreased in recent years, possibly due to stricter Australian visa regulations and/or the perceived inadequacy of the British pension for the Australian costs of living. A newspaper report ('Bid to unfreeze pensions of retired British expats', *The Australian*, 28 February 2005, p. 5) describes the British government practice of refusing to index its overseas pensions against inflation. As a result, British pensioners living in Australia rapidly lose the real value of their pension.
23. Sometimes elderly migrants from a specific language and ethnic background are placed together in a so-called 'cluster' within a general hostel or nursing home.
24. British subjects who were on a Commonwealth electoral roll as at 25 January 1984 are eligible to vote.
25. Interviewee 31181.
26. Interviewee 31172.
27. Interviewee 31072.
28. Interviewee 30081.
29. Several Irish migrants in fact had three passports: Irish, Australian and British.
30. There was only one exception, an Irish migrant who was concerned that changes in the political climate of Australia (brought about at the time by the ascendency of the One Nation Party) might lead to some migrants being 'thrown out'.
31. For example, *The Australian* (2 March 2005, Features, p. 11) has an extended account of the deportation of a Tongan family.

8 Conclusion: Towards a model of transnational caregiving

1. As discussed in Chapter 4, reciprocity is the most important principle in the process of negotiation regarding the fulfilment of family obligations to ensure that people feel there is no imbalance in the caregiving relations (Finch & Mason, 1993; Grundy, 2005). In anthropology, generalised reciprocity is defined as a form of exchange in which neither the time nor the value of the return are specified. Finch and Mason (1993) extend this meaning to caregiving exchanges that do not necessarily involve the same people giving and receiving care, as when people feel an obligation to pay back to a third party, even to the community. The reciprocal gift may come much later or be given by another person.
2. Of course, there are many transnational contexts that do not involve significant distances, for example, when kin live either side of a nation-state border.
3. For the purposes of this study, we do not include in our definition of transnationals people who have a close affinity to another country, without having actual kinship ties. Some people identify with another nation-state without having any family, kin or friendship connections in that place. Such people might define themselves, for example, as 'Francophiles', or speak of

England as 'home', see themselves as 'citizens of the world' or as one person described herself as a 'displaced European'. Perhaps it could be said that these people *think* transnationally, but they do not necessarily *practice* transnational kinship relations.

4. A common example of the activation of dormant transnational kin relations in the migration literature is of aspiring migrants who trace kinship connections to individuals in their destination of choice by utilising their existing active kinship networks until they find someone through whom they can make contact with a relative living abroad (e.g., di Leonardo, 1987). In this way, the aspiring migrants hope to activate ties that might assist them in their migration.
5. The financial obligations and expectations of migrants have been discussed at length in the literature on so called MIRAB (migration, remittances, aid and bureaucracy) economies, particularly of the small Pacific Islands (Lee, 2003; Bertram, 2004).
6. Hannerz (1992, p. 252) defines transnationals as frequent travellers (usually occupational) who share 'structures of meaning carried by social networks' (pp. 248–249, see also Werbner, 1999, p. 17).
7. A similar review article of caretaking from an anthropological perspective by Jennie Keith (1992) also makes no mention of transnational contexts.
8. Williams (2004) makes reference to 'transnational kinship' (p. 7) as one of five central research projects developed at the CAVA Research Programme on Care, Values and the Future of Welfare that informs the findings of her book. She also refers to 'transnational kin' (p. 55) in the context of changing caregiving relationships. However, there is no sustained consideration or focus on transnational caregiving in the book.
9. The concern about dual citizenship has been further heightened in very recent times by the perceived threat of 'home-grown terrorists', second-generation migrants who were born and educated in the countries to which they are deemed to pose a potential terrorist threat.
10. Werbner (1997) uses the term 'translocals' to refer to the proletarian version of transnationals. We reserve this term for kin who are separated by distance but within nation-state borders.

References

ABS (2001) *Australian Bureau of Statistics 2001 Census*, available from http://www.abs.gov.au [accessed September 2005].

ABS (2004) *Disability, Ageing and Carers: Summary of Findings, Australia 2003*. Australian Bureau of Statistics, Canberra, ABS Cat. No. 4430.0.

ABS (2005) *2005 Year Book Australia*. Cat. No. 1301.0, No. 87. Australian Bureau of Statistics, Canberra.

Ackers, L. (1998) *Shifting Spaces: Women, Citizenship and Migration within the European Union*. Policy Press, Bristol.

Ackers, L. & Dwyer, P. (2002) *Senior Citizenship? Retirement, Migration and Welfare in the European Union*. Policy Press, Bristol.

Ackers, L. & Stalford, H. (2004) *A Community for Children? Children, Citizenship and Internal Migration in the EU*. Ashgate, Aldershot.

Aldous, J., Klaus, E. & Klein, D. (1985) 'The understanding heart: Ageing parents and their favourite children', *Child Development*, 56, 303–316.

Andall, J. (1999) 'Cape Verdean women on the move: "Immigration shopping" in Italy and Europe', *Modern Italy*, vol. 4, no. 2, pp. 241–257.

Anderson, B. & Tracey, K. (2001) 'Digital living: The impact (or otherwise) of the internet on everyday life', *The American Behavioral Scientist*, vol. 45, no. 3, pp. 456–475.

Anderson, G. & Hussey, P. (2000) 'Population aging: A comparison among industrialized countries', *Health Affairs*, vol. 19, no. 3, pp. 191–204.

Ang, I. (1994) 'On not speaking Chinese: Postmodern ethnicity and the politics of diaspora', *New Foundations*, vol. 24, pp. 1–18.

Appadurai, A. (1991) 'Global ethnoscapes: Notes and queries for a transnational anthropology', in R. Fox (ed.), *Recapturing Anthropology*, School of American Research Press, Santa Fe, New Mexico, pp. 191–210.

Appleyard, R. & Baldassar, L. (2004) 'Peopling Western Australia', in R. Wilding & F. Tilbury (eds) *A Changing People: Diverse Contributions to the State of Western Australia*, Department of Premier and Cabinet, Perth Western Australia, pp. 8–32.

Arber, S. & Evandrou, M. (eds) (1993) *Ageing, Independence and the Life Course*. Kingsley Publishers, London.

Arensberg, C.M. (1937) *The Irish Countryman: An Anthropological Study*. Macmillan, London.

Ashton, T. (2000) 'New Zealand: Long-term care in a decade of change', *Health Affairs*, vol. 19, no. 3, pp. 72–86.

Australian Citizenship Council (1999) *Contemporary Australian Citizenship*. DIMA, Canberra.

Baldassar, L. (1992) 'Italo-Australian youth in Perth: Space speaks and clothes communicate', in R. Bosworth & R. Ugolini (eds), *War, Internment and Mass Migration: The Italo-Australian Experience 1940–1990*, Gruppo Editorial Internazionale, Rome, pp. 207–224.

Baldassar, L. (1998) 'The return visit as pilgrimage: Secular redemption and cultural renewal in the migration process', in E. Richards & J. Templeton (eds), *The Australian Immigrant in the Twentieth Century: Searching Neglected Sources*, Division of Historical Studies and Centre for Immigration and Multicultural Studies, Research School of Social Sciences, ANU Canberra, pp. 127–156.

Baldassar, L. (2001) *Visits Home: Migration Experiences between Italy and Australia*. Melbourne University Press, Melbourne.

Baldassar, L. (2004a) 'Italians in Western Australia: From "dirty dings" to multicultural mates', in R. Wilding & F. Tilbury (eds), *A Changing People: Diverse Contributions to the State of Western Australia*, Department of Premier and Cabinet, Perth WA, pp. 266–283.

Baldassar, L. (2004b) 'Italians in Australia', in M. Ember, C.R. Ember & I. Skoggard (eds), *Encyclopedia of Diasporas: Immigrant and Refugee Cultures around the World*. Kluwer Academic/Plenum Publishers, New York.

Baldassar, L. (2007, in press) 'Transnational families and aged care: The mobility of care and the migrancy of ageing', *Journal of Ethnic and Migration Studies*, vol. 33, no. 4.

Baldassar, L. & Baldock, C. (2000) 'Linking migration and family studies: Transnational migrants and the care of ageing parents', in B. Agozino (ed.) *Theoretical and Methodological Issues in Migration Research*. Ashgate, Aldershot, pp. 61–89.

Baldassar, L., Baldock, C. & Lange, C. (1999) 'Immigration and transnational care-giving: Public policies and their impact on migrants' ability to care from a distance', in M. Collis (ed.) *Challenges and Prospects: Sociology for a New Millenium, TASA Conference Proceedings*, CeLTS, Melbourne, pp. 465–474.

Baldassar, L. & Pesman, R. (2005) *From Paesani to Global Veneti: Veneto migrants in Australia*. UWA Press, Crawley.

Baldassar, L., Wilding, R. & Baldock, C. (2006, in press) 'Long-distance care-giving: Transnational families and the provision of aged care', in I. Paoletti (ed.) *Family Caregiving for Older Disabled People: Relations and Institutional Issues*. Nova Science Publishers, New York.

Baldock, C.V. (1999) 'The ache of frequent farewells', in S. Feldman & M. Poole (eds) *A Certain Age: Women Growing Older*. Allen & Unwin, Sydney, pp. 182–192.

Baldock, C.V. (2000) 'Migrants and their parents: Caregiving from a distance', *Journal of Family Issues*, vol. 21, no. 2, pp. 205–224.

Baldock, C., Baldassar, L. & Lange, C. (1999) 'Immigrants as long-distance carers', in M. Pember (ed.) *Post Haste the Millenium: Opportunities and Challenges in Local Studies*, Proceedings of the 2nd National Australian Library and Information Association Ltd Local Studies Section Conference, Local Studies Section, ALIA, Perth, pp. 213–222.

Barr, M.D. (2000) 'Lee Kuan Yew and the "Asian values" debate', *Asian Studies Review*, vol. 24, no. 3, pp. 309–334.

Barr, M.D. (2002) *Cultural Politics and Asian Values: The Tepid War*. Routledge, London.

Basch, L., Glick Schiller, N. & Szanton Blanc, C. (1994) *Nations Unbound: Transnational Projects, Postcolonial Predicaments, and Deterritorialized Nation-States*. Gordon and Breach, Langhorne, PA.

Batrouney, T. & Stone, W. (1998) 'Cultural diversity and family exchanges', *Family Matters*, no. 51, pp. 13–20.

Bauböck, R. (1994) *Transnational Citizenship: Membership and Rights in International Migration*. Edward Elgar, Aldershot.

Beck, U. (2002) 'The cosmopolitan perspective: Sociology of the second age of modernity', in S. Vertovec & R. Cohen (eds) *Conceiving Cosmopolitanism: Theory, Context, Practice*. Oxford University Press, Oxford.

Bengtson, V.L. & Achenbaum, W.A. (eds) (1993) *The Changing Contract across the Generations*. Aldine de Gruyter, New York.

Bengtson, V.L., & Harootyan, R.A., Kronebusch, K., Lawton, L., Schlesinger, M., Silverstein, M. & Vorek, R.E. (1994) *Intergenerational Linkages: Hidden Connections in American Society*. Springer, New York.

Bengtson, V.L., Kim, K., Myers, G.C., & Eun, K. (eds) (2000) *Aging in East and West: Families, States, and the Elderly*. Springer Publishing Company, New York.

Bengtson, V.L. & Roberts, R.E.L. (1991) 'Intergenerational solidarity in aging families: An example of formal theory construction', *Journal of Marriage and the Family*, vol. 53, no. 4, pp. 856–870.

Bengtson, V., Rosenthal, C. & Burton, L. (1995) 'Paradoxes of family and aging', in R.H. Binstock & L.K. George (eds) *Handbook of Aging and the Social Sciences*, 4th edn. Academic Press, San Diego, pp. 253–282.

Benton, G. (2003) 'Chinese transnationalism in Britain: A longer history', *Identities: Global Studies in Power and Culture*, vol. 10, no. 3, pp. 347–375.

Bertelli, L. (1985) 'Italian Families', in D. Storer (ed.), *Ethnic Family Values in Australia*. Prentice Hall, Sydney.

Bertram, G. (2004) 'The Mirab model in the 21st century', in *Changing Islands – Changing Worlds*, Islands of the World VIII International Conference, 1–7 November, 2004, Kinment Island, Taiwan.

Betteridge, J. (1997) 'Answering back: The telephone, modernity and everyday life', *Media, Culture & Society*, vol. 19, pp. 585–603.

Bettoni, C. & Rubino, A. (1996) *Emigrazione e Comportamento Linguistico: un 'indagine sul trilinguismo dei Siciliani e dei Veneti in Australia*, Sociolinguistica e Dialettologia. Collana diretta da A. Sobrero, vol. 7, Congedo Editoriale, Galatina (Le).

Bhabha, H. (1989) 'Location, intervention, incommensurabliity: A conversation with Homi Bhabha', *Emergences*, vol. 1, no. 1, pp. 63–88.

Bijsterveld, K. (1996) *Geen kwestie van Leeftijd: Verzorgingsstaat, wetenschap en discussies rond ouderen in Nederland, 1945–1882*. Van Gennep, Amsterdam.

Bittman, M., Fast, J., Fisher, K. & Thomson, C. (2004) 'Making the invisible visible: The life and time(s) of informal caregivers', in N. Folbre & M. Bittman (eds) *Family Time: The Social Organization of Care*, Routledge, London, pp. 69–89.

Blackman, T. (2000) 'Defining responsibility for care: Approaches to the care of older people in six European countries', *International Journal of Social Welfare*, vol. 9, no. 3, pp. 181–190.

Blackman, T., Brodhurst, S. & Convery, J. (eds) (2001) *Social Care and Social Exclusion: A Comparative Study of Older People's Care in Europe*. Houndmills, Basingstoke, Hampshire.

Blainey, G. (1966) *The Tyranny of Distance: How Distance Shaped Australia's History*. Sun Books, Melbourne.

Blanc, C.S., Basch, L. & Schiller, N.G. (1995) 'Transnationalism, nation-states and culture', *Current Anthropology*, vol. 36, no. 4, pp. 683–686.

Bonafazi, C. (1998) *L'Immigrazione straniera in Italia*. Il Mulino, Bologna.

Borrie, W. (1994) *The European Peopling of Australasia: A Demographic History*. Demography Program Australian National University, Canberra.

Bosworth, R, (1996) *Italy and the Wider World*. Routledge, New York.

Bottomley, G. (1992) *From Another Place: Migration and the Politics of Culture*. Cambridge University Press, Cambridge.

Boyd, S. (2006, in press) 'Communication and community: Perspectives on language policy in Sweden and Australia since the mid 1970s', in A. Pauwels (ed.) *Maintaining Minority Languages in Transnational World*. Palgrave Macmillan, Houndmills, Basingstoke, Hampshire.

Brosnan, P. & Poot, J. (1986) *An Econometric Model of Trans-Tasman Migration After World War II*. The Australian National University Centre for Economic Policy Research, Canberra.

Bryceson, D. & Vuorela, U. (2002) *The Transnational Family: Global European Networks and New Frontiers*. Berg, Oxford.

Buetow, S. (1994) 'International migration: Some consequences for urban areas in Australia and New Zealand'. *International Migration*, vol. 32, no. 2, pp. 307–325.

Burn, J. & Reich, S. (2005) *The Immigration Kit: A Practical Guide to Australia's Immigration Law*, 7th edn. The Federation Press, Sydney.

Cairncross, F. (1997) *The Death of Distance: How the Communications Revolution will Change Our Lives*. Harvard Business School Press, Boston.

Carmichael, G. (1993) *A History of Migration between New Zealand and Australia*. Australian National University Working Papers in Demography, Canberra.

Carmichael, G. (1996) 'Trans-Tasman migration', in P. Newton & M. Bell (eds) *Population Shift: Mobility and Change in Australia*, AGPS, Canberra, pp. 39–57.

Carmichael, G. (2001) 'New Zealanders', in J. Jupp (ed.) *The Australian People: An Encyclopedia of the Nation, Its People and Their Origins*, 2nd edn. Cambridge University Press, pp. 602–608.

Castles, S. (1997) 'Multicultural citizenship: A response to the dilemma of globalisation and national identity?', *Journal of Intercultural Studies*, vol. 18, no. 1, pp. 5–22.

Castles, S., Alcorso, C., Rando, G. & Vasta, E. (eds) (1992) *Australia's Italians: Culture and Community in a Changing Society*. Allen & Unwin, Sydney.

Cheah, P. & Robbins, B. (eds) (1998) *Cosmopolitics: Thinking and Feeling Beyond the Nation*. University of Minnesota Press, Minneapolis.

Chetkovich, J. (2002) *The 'New Irish' in Australia: A Western Australian Perspective*. History PhD Thesis, University of Western Australia, Perth.

Chetkovich, J. (2004) 'There would seem to be a wonderful freedom out here: The Irish in Western Australia', in R. Wilding & F. Tilbury (eds), *A Changing People: Diverse Contributions to the State of Western Australia*, Department of Premier and Cabinet, Perth WA, pp. 222–235.

Chiro, G. & Smolicz, J. (1998) 'Evaluations of language and social systems by a group of tertiary students of Italian ancestry in Australia', vol. 18, pp. 13–31.

Chiro, G. & Smolicz, J. (2002) 'Italian family values and ethnic identity in Australian schools', *Educational Practice and Theory*, vol. 24, no. 2, pp. 37–51.

Cicirelli, V.G. (1995) *Sibling Relationships across the Life Span*. Plenum, New York.

Climo, J. (1992) *Distant Parents*. Rutgers University Press, New Brunswick.

Clyne, M. (1972) *Perspectives on Language Contact*. Hawthorn Press, Melbourne.
Clyne, M., & Kipp, S. (2002) 'Australia's changing language demography', *People and Place*, vol. 10, no. 3, pp. 29–35.
Cohen, E. (1972) 'Towards a sociology of international tourism', *Social Research*, vol. 39, no. 1, pp. 164–182.
Cohen, R. (1997) *Global Diasporas: An Introduction*. University of Wellington Press, Seattle.
Commonwealth of Australia (1989) *National Agenda for a Multicultural Australia*.
Contessa, E. (2000) 'Care of the Italian aged', in P. Genovesi & I. Musolino (eds), *I.A.I Italian Australian Institute in Search of the Italian Australian into the New Millennium Conference Proceedings*, Gro-Set Pty Ltd, Thornbury, Victoria, pp. 489–491.
Crock, M. (1998) *Immigration and Refugee Law in Australia*. Federation Press, Leichhardt.
Curry, J. (1998) *Irish Social Services*. Institute of Public Administration, Dublin.
Daly, M. (1998) 'A more caring state? The implications of welfare state restructuring for social care in the Republic of Ireland', in J. Lewis (ed.), *Gender, Social Care and Welfare State Restructuring in Europe*. Ashgate, Aldershot, pp. 25–50.
Delaney, C. (1990) 'The Hajj: Sacred and secular', *American Ethnologist*, vol. 17, pp. 513–530.
Department of Immigration and Multicultural Affairs, April 1999, *Fact Sheet 29*.
di Leonardo, M. (1987) 'The female world of cards and holidays: Women, families, and the work of kinship', *Signs*, vol. 12, no. 3, pp. 440–453.
Dilworth-Anderson, P., Williams, I.C. & Gibson, B.E. (2002) 'Issues of race, ethnicity and culture in caregiving research: A 20-year review (1980–2000)', *The Gerontologist*, vol. 42, no. 3, pp. 237–272.
DIMA (Department of Immigration and Multicultural Affairs) (1997) *New Zealanders in Australia*. DIMA, Canberra.
Dimaggio, P., Hargittai, E., Neuman, W.R. & Robinson, J.P. (2001) 'Social implications of the Internet', *Annual Review of Sociology*, vol. 27, pp. 307–336.
Doty, P. (1995) 'Older caregivers and the future of informal caregiving', in S.A. Bass (ed.) *Older and Active: How Americans over 55 are Contributing to Society*. Yale University Press, New Haven, pp. 96–121.
Duff, J. (2001) 'Financing to foster community health care: A comparative analysis of Singapore, Europe, North America and Australia', *Current Sociology*, vol. 49, no. 3, pp. 135–154.
Duval, D.T. (2003) 'When hosts become guests: Return visits and diasporic identities in a Commonwealth Eastern Caribbean community', *Current Issues in Tourism*, vol. 6, no. 4, pp. 267–308.
Duval, D.T. (2004) 'Linking return visits and return migration among Commonwealth Eastern Caribbean migrants in Toronto', *Global Networks*, vol. 4, no. 1, pp. 510–567.
Eriksen, T.H. (1991) 'The cultural contexts of ethnic differences', *Man* (N.S.), vol. 26, pp. 127–144.
Extra, G. (2006, in press) 'Comparative perspectives on immigrant minority languages in multicultural Europe', in A. Pauwels (ed.) *Maintaining Minority Languages in Transnational World*, Palgrave Macmillan, Houndmills, Basingstoke, Hampshire.

Extra, G., Aarts, R., van der Avoird, T. & Yağmur, K. (2002) *De andere talen van Nederland*. Bussum, Countinho.

Fergusson, D., Hong, B., Horwood, J., Jensen, J. & Travers, P. (2001) *Living Standards of Older New Zealanders: A Summary*. The Ministry of Social Policy, Wellington.

Finch, J. (1989) *Family Obligations and Social Change*. Cambridge, Polity Press.

Finch, J. & Groves, D. (1983) *A Labour of Love: Women, Work and Caring*. Routledge & Kegan Paul, London.

Finch, J. & Mason, J. (1990) 'Divorce, remarriage and family obligations'. *Sociological Review*, vol. 38, no. 2, pp. 219–246.

Finch, J. & Mason, J. (1991) 'Obligations of kinship in contemporary Britain: Is there normative agreement?', *British Journal of Sociology*, vol. 42, no. 3, pp. 344–367.

Finch, J. & Mason, J. (1993) *Negotiating Family Responsibilities*. Routledge, London.

Fine, M. (2004) 'Renewing the social vision of care', *Australian Journal of Social Issues*, vol. 39, no. 3, pp. 217–232.

Fine, M. (2005) 'Individualization, risk and the body: Sociology and care', *Journal of Sociology*, vol. 41, no. 3, pp. 247–266.

Fortier, A. (2003) 'Community, belonging, and the effervescence of ethnicity', paper presented in Italian Diasporas Share the Neighbourhood, Europeans Seminar, Institute of Advanced Studies, University of Western Australia, July 2003, Perth Western Australia.

Fougere, G. (2001) 'Transforming health sectors: New logics of organizing in the New Zealand health system', *Social Science & Medicine*, vol. 52, pp. 1233–1242.

Gabaccia, D. (2000) *Italy's Many Diasporas*. University of Washington Press, Seattle.

Gabaccia, D. & Ottanelli, F.M. (2001) *For Us There are No Frontiers: Labor, Migration and the Making of Multi-Ethnic Nations*. University of Illinois Press, Urbana.

Gardner, K. (2002) 'Death of a migrant: Transnational death rituals and gender among British Sylhetis', *Global Networks*, vol. 2, no. 3, pp. 191–204.

Gardner, K. & Grillo, R. (2002) 'Transnational households and ritual: An overview', *Global Networks*, vol. 2, no. 3, pp. 179–190.

Giarchi, G. (1996) *Caring for Older Europeans: Comparative Studies in 29 Countries*. Ashgate, Aldershot.

Giddens, A. (1991) *Modernity and Self-Identity: Self and Society in the Late Modern Age*. Polity Press, Cambridge.

Gilroy, P. (1994) 'Diaspora', *Paragraph*, vol. 17, no. 3, pp. 207–212.

Glick Schiller, N., Basch, L. & Blanc-Szanton, C. (1992) *Towards a Transnational Perspective on Migration: Race, Class, Ethnicity and Nationalism Reconsidered*. The New York Academy of Sciences, New York.

Goffman, E. (1963) *Behaviour in Public Places*. Free Press, New York.

Goggin, G. & Newell, C. (2003) *Digital Disability: The Social Construction of Disability in New Media*. Littlefield, Lanham MA, pp. 309–311.

Goggin, G. & Newell, C. (eds) (2006) 'Disability, identity, and interdependence: ICTs and new social forms', Special Issue, *Information, Communication and Society*, vol. 9, no. 3.

Grillo, R. (2007, in press) 'Betwixt and between: Trajectories and projects of transmigration', in Harney, N. & Baldassar, L. (eds) *Journal of Ethnic and Migration*

Studies, Special Issue, 'Tracking Transationalism: Migrancy and its Futures', vol. 33, no. 4.

Grundy, E. (2005) 'Reciprocity in relationships: Socio-economic and health influences on intergenerational exchanges between Third Age parents and their adult children in Great Britain', *British Journal of Sociology*, vol. 56, no. 2, pp. 233–255.

Gubrium, J.F. (1997) *Living and Dying at Murray Manor*, 2nd edn. University Press of Virginia, Charlotteville.

Gubrium, J.F. & Holstein, J.A. (1990) *What is Family?* Mayfield, Mountain View.

Guest, P. (1993) 'Distribution and mobility of the New Zealand-born in Australia', in G. Carmichael (ed.) *Trans-Tasman Migration: Trends, Causes and Consequences*, AGPS, Canberra, pp. 216–254.

Gupta, A. & Ferguson, J. (1997) *Anthropological Locations: Boundaries and Grounds of a Field Science*. University of California Press, Berkeley.

Gupta, R. & Pillai, V.K. (2000) 'Elder care giver burden in south Asian families in the Dallas-Fort Worth metropolitan area', *Hallym International Journal of Aging*, vol. 2, no. 2, pp. 93–104.

Hage, G. (2003) *Against Paranoid Nationalism: Searching for Hope in a Shrinking Society*. Pluto Press, Annandale NSW.

Hagemann, R.P. & Nicoletti, G. (1989) *Ageing Populations: Economic Effects and Implications for Public Finance*. OECD Department of Economics and Statistics Working Papers, No. 61. Paris.

Hagestad, G.O. (1996) 'On time, off-time, out of time? Reflections on continuity and discontinuity from an illness process', in V. Bengtson (ed.) *Adulthood and Aging: Research on Continuities and Discontinuities*. Springer Publishing Co., New York, pp. 204–227.

Hall, S. (2002) 'Political Belonging in a World of Multiple Identities' in S. Vertovec & R. Cohen (eds) *Cosmopolitanism: Theory, Context, Practice*. Oxford University Press, Oxford.

Hammersley, M. & Atkinson, P. (1995) *Ethnography: Principles in Practice*, 2nd edn. Routledge, London.

Hannerz, U. (1992) *Cultural Complexity: Studies in the Social Organization of Meaning*. Columbia University Press, New York.

Hareven, T.K. & Adams, K.J. (1982) *Aging and Life Course Transitions: An Interdisciplinary Perspective*. Guilford Press, New York.

Harney, N. & Baldassar, L. (eds) (2007, in press) 'Tracking transnationalism: Migrancy and its futures', Special Issue, *Journal of Ethnic and Migration Studies*, vol. 33, no. 4.

Harvey, D. (2005) *A Brief History of Neoliberalism*. Oxford University Press, Oxford.

Hennessy, P. (1995) *Social Protection for Dependent Elderly People: Perspectives From a Review of OECD Countries*. Labour Market and Social Policy Occasional Paper, No. 16. Paris, OECD.

Horkan, E. (1995) 'Elder abuse in the Republic of Ireland', in J. Kosberg & J. Garcia (eds) *Elder Abuse: International and Cross Cultural Perspectives*. The Haworth Press, New York, pp. 119–137.

ITU (2003) *World Telecommunication Development Report. Access Indicators for the Information Society, Summary*. International Telecommunications Union, World Summit on the Information Society. Avail: http://www.itu.int/ITUD/ict/publications/wtdr_03/material/WTDR2003Sum_e.pdf [Accessed July 2005].

Jayasuriya, L. (1990) 'Rethinking Australian multiculturalism: Towards a new paradigm', *Australian Quarterly*, vol. 62, no. 1, pp. 51–63.

Jones, S. (1983) *An Exploratory Study into Selected Social and Political Attitudes of Perth's Maori Population*. MA Dip. Thesis, Department of Politics, University of Western Australia, Perth.

Joseph, A. & Chalmers, A. (1996) 'Restructuring long-term care and the geography of ageing: A view from rural New Zealand', *Social Science & Medicine*, vol. 42, no. 6, pp. 887–896.

Joseph, A.E. & Hallman, B.C. (1998) 'Over the hill and far away: Distance as a barrier to the provision of assistance to elderly relatives', *Social Science Medicine*, vol. 46, no. 6, pp. 631–639.

Jupp, J. (1998) *Immigration*. Oxford University Press, Melbourne.

Jupp, J. (ed.) (2001) *The Australian People: An Encyclopedia of the Nation, its People and their Origins*, 2nd edn. Cambridge University Press.

Kamalkhani, Z. (2004) 'Migration from Iran, Iraq and Afghanistan: Seeking refuge and building lives', in R. Wilding & F. Tilbury (eds) *A Changing People: Diverse Contributions to the State of Western Australia*, Department of Premier and Cabinet, Perth Western Australia, pp. 236–251.

Kanter, R.M. (1995) *World Class: Thriving Locally in the Global Economy*. Simon & Shuster, New York.

Katz, S. (1996) *Disciplining Old Age: The Formation of Gerontological Knowledge*. University Press of Virginia, Charlottesville and London.

Keating, E. & Mirus, G. (2003) 'American sign language in virtual space', *Language in Society*, vol. 32, no. 5, pp. 693–714.

Keith, J. (1992) 'Care-taking in cultural context: Anthropological queries', in H. Kendig, A. Hashimoto & L. Coppard (eds) *Family Support for the Elderly: The International Experience*. Oxford University Press, Oxford.

Kennedy, F. (1986) 'The family in transition', in K. Kennedy (ed.), *Ireland in Transition: Economic and Social Change Since 1960*. Mercier, Cork, pp. 91–100.

Khoo, S. (1994) 'Correlates of welfare dependency among immigrants in Australia', *International Migration Review*, vol. 28, no. 1, pp. 68–92.

King, R. and Andall, J. (1999) 'The geography and economic sociology of recent immigration to Italy', *Modern Italy*, vol. 4, no. 2, pp. 135–158.

King, R., Warnes, T. & Williams, A. (2000) *Sunset Lives, British Retirement Migration to the Mediterranean*. Berg, New York.

Klymlicka, W. (1995), *Multicultural Citizenship: A Liberal Theory of Minority Rights*. Clarendon Press, Oxford.

Knijn, T. & Kremer, M. (1997) 'Gender and the caring dimension of welfare states: Toward inclusive citizenship', *Social Politics*, vol. 4, no. 3, pp. 328–361.

Koser, K. (1997) 'Social networks and the asylum cycle: The case of Iranians in the Netherlands', *International Migration Review*, vol. 31, no. 3, pp. 591–611.

Kramarae, C. (ed.) (1988) *Technology and Women's Voices: Keeping in Touch*. Routledge & Kegan Paul, New York.

Lacohee, H. & Anderson, B. (2001) 'Interacting with the telephone', *International Journal of Human-Computer Studies*, vol. 54, no. 5, pp. 665–699.

Lamont, M. & Aksartova, S. (2002) 'Ordinary cosmopolitanisms: Strategies for bridging racial boundaries among working-clanss men', *Theory, Culture & Society*, vol. 19, no. 4, pp. 1–25.

Larragy, J. (1993a) 'Views and perceptions of older Irish people', *Social Policy and Administration*, vol. 27, no. 3, pp. 235–247.

Larragy, J. (1993b) 'Formal service provision and the care of the elderly at home in Ireland', *Journal of Cross-Cultural Gerontology*, vol. 8, pp. 361–374.

Leane, M. (1995) 'Female carers for elderly people: Implications for women's labour market participation in Ireland', in J. Phillips (ed.) *Working Carers: Avebury Studies of Care in the Community*, Avebury, Ashgate, Aldershot, pp. 58–72.

Lee, H. (2003) *Tongans Overseas: Between Two Shores*. University of Hawaii Press, Honolulu.

Lee, E.F. & Yeo, Y.F. (2003) 'Singapore's demographic trends in 2002', *Statistics Singapore Newsletter*, September 2003.

Licoppe, C. (2004) '"Connected" presence: The emergence of a new repertoire for managing social relationships in a changing communication technoscape', *Environment and Planning D: Society and Space*, vol. 22, pp. 135–156.

Lin, G. & Rogerson, P.A. (1995) 'Elderly parents and the geographic availability of their adult children', *Research on Aging*, vol. 17, pp. 303–331.

Litwak, E. & Kulis, S. (1987) 'Technology, proximity and measures of kin support'. *Journal of Marriage and the Family*, vol. 49, pp. 649–661.

Loizos, P. (2000) 'Are refugees social capitalists?', in S. Baron, J. Field & T. Schuller (eds) *Social Capital: Critical Perspectives*. Oxford University Press, Oxford.

MacDonagh, O. (1996) *The Sharing of the Green: A Modern Irish History for Australians*. Allen & Unwin, St Leonards.

MacKinnon, V.J. (1998) 'Language difficulties and health consequences for older Italian-Australians in Ascot Vale', *Australian Journal of Primary Health*, vol. 4, no. 4, pp. 31–43.

MacKinnon, V. & Nelli, A. (1996) *Now We are in Paradise, Everything is Missing*. Department of Nursing, Victoria University of Technology, Footscray.

Mahler, S.J. (1998) 'Theoretical and empirical contributions toward a research agenda for transnationalism', in L.E. Guarnizo & M.P. Smith (eds) *Transnationalism from Below*. Transaction Publishers, New Brunswick, NJ, pp. 64–100.

Mandel, R. (1990) 'Shifting centres and emergent identities: Turkey and Germany in the lives of Turkish gastarbeiter', in D. Eichelman & J. Piscatori (eds), *Muslim Travellers: Pilgrimage, Migration and the Religious Imagination*. Routledge, London, pp. 153–171.

Manderson, L. (1990) 'Introduction: Does culture matter?', in J. Reid & P. Trompf (eds) *The Health of Immigrant Australia: A Social Perspective*. Harcourt Brace Jovanovich, Sydney, pp. xi–xvii.

Marcus, G. (1995) 'Ethnography in/of the world system: The emergence of multi-sited ethnography', *Annual Review of Anthropology*, vol. 24, pp. 95–117.

Mason, J. (1999) 'Living away from relatives: Kinship and geographical reasoning', in S. McRae (ed.) *Changing Britain: Families and Households in the 1990s*. Oxford University Press, Oxford, pp. 156–175.

Mason, J. (2004) 'Managing kinship over long distances: The significance of "the visit"', *Social Policy and Society*, vol. 3, no. 4, pp. 421–429.

Matthews, S. & Rosner, T. (1988) 'Shared filial responsibility: The family as the primary caregiver', *Journal of Marriage and the Family*, vol. 50, pp. 185–195.

Merlis, M. (2000) 'Caring for the frail elderly: An international review', *Health Affairs*, vol. 19, no. 3, pp. 141–150.

Miller, D. & Slater, D. (2000) *The Internet: An Ethnographic Approach*. Berg, Oxford.

Mitchell, J.C. (1959) 'The causes of labour migration', *Bulletin of the Inter-African Labour Institute*, vol. 6, no. 1, pp. 12–44.

Moss, M.S. & Moss, S.Z. (1992) 'Themes in parent-child relationships when elderly parents move nearby', *Journal of Aging Studies*, vol. 6, no. 3, pp. 259–271.

Na, M. (2001) 'The home computer in Korea: Gender, technology, and the family'. *Feminist Media Studies*, vol. 1, no. 3, pp. 291–306.

NCAOP (2001a) *Demography: Ageing in Ireland Fact File No. 1*. National Council on Ageing and Older People, Dublin.

NCAOP (2001b) *Physical Health: Ageing in Ireland Fact File No. 2*. National Council on Ageing and Older People, Dublin.

NCAOP (2001c) *Income: Ageing in Ireland Fact File No. 3*. National Council on Ageing and Older People, Dublin.

Nelde, P.H. (2006, in press) 'Propositions for a European language policy', in A. Pauwels (ed.) *Maintaining Minority Languages in Transnational World*. Palgrave Macmillan, Houndmills, Basingstoke, Hampshire.

Ng, A.C.Y., Phillips, D.R. & Lee, W.K. (2002) 'Persistence and challenges to filial piety and informal support of older persons in a modern Chinese society: A case study in Tuen Mun, Hong Kong', *Journal of Aging Studies*, vol. 16, pp. 135–153.

Ng, S.H. & McCreanor, T. (1999) 'Patterns in discourse about elderly people in New Zealand', *Journal of Aging Studies*, vol. 13, no. 4, pp. 473–489.

Nydegger, C. (1991) 'The development of paternal and filial maturity', in K. Pillemer & K. McCarthney (eds) *Parent-Child Relations throughout Life*. Lawrence Erlbaum, Hillsdale, NJ, pp. 93–112.

O'Connor, D. (1996) *No Need to be Afraid. Italian Settlers in South Australia between 1839 and the Second World War*. Wakefield Press, Adelaide.

OECD (1988) *Ageing Populations: The Social Policy Implications*. OECD, Paris.

O'Farrell, P. (2001) *The Irish in Australia: 1788 to the Present*. University of Notre Dame Press, Indiana.

Olwig, K.F. (2002) 'A wedding in the family: Home making in a global kin network'. *Global Networks*, vol. 2, no. 3, pp. 205–218.

Olwig, K.F. (2003) '"Transnational" socio-cultural systems and ethnographic research: Views from an extended field site', *International Migration Review*, vol. 37, no. 3, pp. 787–811.

Opie, A. (1992) *There's Nobody there: Community Care of Confused Older People*. Oxford University Press, Auckland.

Opie, A. (1994) 'The instability of the caring body: Gender and caregivers of confused older people', *Qualitative Health Research*, vol. 4, no. 1, pp. 31–50.

Ortner, S. (1997) 'Fieldwork in the postcommunity', *Anthropology and Humanism*, vol. 22, no. 1, pp. 61–80.

Ostergaard-Nielsen, E. (ed.) (2003) *International Migration and Sending Countries: Perceptions, Policies and Transnational Relations*. Palgrave Macmillan, London.

Pahl, J. (1989) *Money and Marriage*. Macmillan, London.

Pahl, J. (1995) 'His money, her money: Recent research on financial organisation in marriage', *Journal of Economic Psychology*, vol. 16, pp. 361–376.

Pahl, J. (2001a) 'Couples and their money: Theory and practice in personal finances', in R. Sykes, C. Bochel & N. Ellison (eds) *Social Policy Review*, 13. Policy Press, Bristol, pp. 17–37.

Pahl, J. (2001b) 'Widening the scope of social policy: Families, financial services and the impact of technology', in R. Edwards & J. Glover (eds) *Risk and Citizenship: Key Issues in Welfare*. Routledge, London, pp. 64–79.

Parsons, T. (1965) 'The normal American family', in S.M. Farber (ed.) *Man and Civilisation*. McGraw-Hill, New York, pp. 31–50.

Pauwels, A. (ed.) (2006, in press) *Maintaining Minority Languages in Transnational World*. Palgrave Macmillan, Houndmills, Basingstoke, Hampshire.

Peters, N. (2004) 'Going Dutch: Four centuries of contact with Netherlanders', in R. Wilding & F. Tilbury (eds), *A Changing People: Diverse Contributions to the State of Western Australia*, Department of Premier and Cabinet, Perth WA, pp. 93–107.

Pijl, M. (2004) 'Caring from a distance', unpublished paper, Den Hague, March 2004.

Pocock, B. (2003) *The Work/Life Collision: What Work is Doing to Australians and What to Do about it*. The Federation Press, Annandale.

Pool, I. de S. (ed.) (1977) *The Social Impact of the Telephone*. MIT Press, Cambridge, MA.

Pool, I. (1980) 'Trans-Tasman migration: An overview', in I. Pool (ed.), *Trans-Tasman Migration*, Population Studies Centre, University of Waikato, New Zealand, Hamilton, pp. 1–20.

Pratt, C. (1980) 'Some characteristics of trans-Tasman travellers, 1976–78: An analysis of Australian arrival and departure data', in I. Pool (ed.) *Trans-Tasman Migration*, Population Studies Centre, University of Waikato, Hamilton, pp. 105–140.

Price, C. (1980) 'Trans-Tasman migration: An Australian viewpoint', in I. Pool (ed.) *Trans-Tasman Migration*, Population Studies Centre, University of Waikato, Hamilton, pp. 66–104.

Quigley, A. & Menon, P. (2004) 'The journey of a thousand miles begins with a single step: Professional migrants from Singapore', in R. Wilding & F. Tilbury (eds) *A Changing People*, Department of Premier and Cabinet, Perth Western Australia, pp. 252–265.

Rosenblatt, B. & Van Steenberg, C. (2003) *Handbook for Long-Distance Caregivers*. Family Caregiver Alliance, San Francisco.

Rosoli, G. (1978) *Un secolo di emigrazione italiana 1876–1976*. Centro Studi Emigrazione, Rome.

Rossi, A.S. & Rossi, P.H. (1990) *Of Human Bonding: Parent–Child Relations Across the Life Course*. Aldine de Gruyter, New York.

Rouse, R. (1991) 'Mexican migration and the social space of postmodernism', *Diaspora*, vol. 1, no. 1, pp. 8–23.

Rowland, D.T. (1991) *Ageing in Australia*. Longman Cheshire, Melbourne.

Rubino, A. (2000) 'Italians and emigration', in G. Moliterno (ed.), *Encyclopedia of Contemporary Italian Culture*. Routledge, London.

Schama, S. (1987) *The Embarrassment of Riches*. Vintage, New York.

Scheper-Hughes, N. (1979) 'Breeding breaks out in the eye of the cat: Sex roles, birth order and the Irish double-bind', *Journal of Comparative Family Studies*, vol. 10, no. 2, pp. 207–226.

Scheper-Hughes, N. (2004) 'Parts unknown: Undercover ethnography of the organs-trafficking underworld', *Ethnography*, vol. 5, no. 1, pp. 29–73.

Schiller, N.G. & Fouron, G. (2001) *Georges Woke up Laughing: Long-Distance Nationalism and the Search for Home*. Duke University Press, Durham.

Schutz, A. (1967) *The Phenomenology of the Social World*. Northwestern University Press, Chicago.

Singh, S. (1997) *Marriage Money: The Social Shaping of Money in Marriage and Banking*. Allen & Unwin, St Leonards.

Skrbis, Z. (1999) *Long Distance Nationalism: Diasporas, Homelands and Identities*. Ashgate, Aldershot.

Skrbis, Z., Kendall, G. & Woodward, I. (2004) 'Locating cosmopolitanism: Between humanist ideal and grounded social category', *Theory, Culture and Society*, vol. 21, no. 6, pp. 115–136.

Smits, J. (1999) 'Family migration and the labour force participation of married women in the Netherlands, 1977–1996', *International Journal of Population Geography*, vol. 5, pp. 133–150.

Smits, J. (2001) 'Career migration, self-selection and the earnings of married men and women in the Netherlands, 1981–93', *Urban Studies*, vol. 38, no. 3, pp. 541–562.

Smolicz, J.J. (1983) 'Modification and maintenance of Italian culture among Italian-Australian youth', *Studi emigrazione*, vol. 20, no. 69, pp. 81–104.

Smoreda, Z. & Licoppe, C. (2000) 'Gender-specific use of the domestic telephone', *Social Psychology Quarterly*, vol. 63, no. 3, pp. 238–252.

Spencer-Cingöz, J. (1998) *Caring for Someone at a Distance*. Age Concern England, London.

Stafford, L. (2005) *Maintaining Long-Distance and Cross-Residential Relationships*. Lawrence Erlbaum Associates, Mahwah, NJ.

Statistics New Zealand (2001) *2001 Census of Population and Dwellings*. Available online: http://www.stats.govt.nz [accessed June 2006].

Statistics New Zealand (2004) *Older New Zealanders – 65 and Beyond*. Available online: http://www.stats.govt.nz [accessed June 2006].

Stein, C.H., Wemmerus, V.A., Ward, M., Gaines, M.E., Freeberg, A.L. & Jewell, T.C. (1998) '"Because they're my parents": An intergenerational study of felt obligation and parental caregiving', *Journal of Marriage and the Family*, vol. 60, no. 3, pp. 611–622.

St John, S. & Willmore, L. (2001) 'Two legs are better than three: New Zealand as a model for old age pensions', *World Development*, vol. 29, no. 8, pp. 1291–1305.

Strand, A., Suhrke, A. & Harpviken, B. (2004) *Afghan Refugees in Iran: From Refugee Emergency to Migration Management*, International Peace Research Institute, Oslo, 16 June 2004.

Sullivan, G. & Gunasekaran, S. (1994) *Motivations of Migrants from Singapore to Australia*. Field Report Series No. 28, Institute of South East Asian Studies.

Templeton, J. (2003) *From the Mountains to the Bush: Italian Migrants Write Home from Australia, 1860–1962*. University of Western Australia Press, Perth WA.

Thane, P. (1998) 'The family lives of old people', in P. Johnson & P. Thane (eds), *Old Age from Antiquity to Post-Modernity*. Routledge, London, pp. 180–210.

Timmermans, J.M. (ed.) (1996) *Rapportage Ouderen 1996*. Sociaal en Cultureel Planbureau, Rijswijk.

Timmermans, J.M., Heide, F., de Klerk, M.M.Y., Kooiker, S.E., Ras, M. & van Dugteren, F.A. (1997) *Vraagverkenning wonen en zorg voor ouderen.* Sociaal en Cultureel Planbureau, Rijswijk.

Torpey, J. (2000) *The Invention of the Passport: Surveillance, Citizenship and the State.* Cambridge University Press, Cambridge.

Treas, J. & Wang, W. (1993) 'Of deeds and contracts: Filial piety perceived in contemporary Shanghai', in V.L. Bengtson & A. Achenbaum (eds), *The Changing Contract Across Generations.* Aldine de Gruyter, New York, pp. 87–98.

Trifiletti, R. (1998) 'Restructuring social care in Italy', in J. Lewis (ed.), *Gender, Social Care and Welfare State Restructuring in Europe.* Ashgate, Aldershot, pp. 175–206.

Ungerson, C. (ed.) (1990) *Gender and Caring: Work and Welfare in Britain and Scandinavia.* Harvester Wheatsheaf, New York.

Urry, J. (1990) *The Tourist Gaze.* Sage, London.

Urry, J. (2000) *Sociology Beyond Societies: Mobilities for the Twenty-First Century.* Routledge, London.

Urry, J. (2002) 'Mobility and proximity', *Sociology,* vol. 36, no. 2, pp. 255–274.

Urry, J. (2003) 'Social networks, travel and talk', *British Journal of Sociology,* vol. 54, no. 2, pp. 155–175.

Urry, J. (2004) 'Connections', *Environment and Planning D: Society and Space,* vol. 22, pp. 27–37.

Uttley, S. (1995) 'New Zealand and community care for older people: A demographic window of opportunity', in T. Scharf & C. Wenger (eds) *International Perspectives on Community Care for Older People.* Avebury, Aldershot, pp. 171–189.

Verdery, K. (1994) 'Ethnicity, nationalism, and state-making: Ethnic groups and boundaries, past and future', in H. Vermeulen & C. Govers (eds) *The Anthropology of Ethnicity: Beyond 'Ethnic Groups and Boundaries'.* Het Spinuis, Amsterdam, pp. 33–57.

Vertovec, S. and Cohen, R. (2002) *Conceiving Cosmopolitanism: Theory, Context, Practice.* Oxford University Press, Oxford.

Ward, C., Bochner, S. & Furnham, A. (2001) 'Refugees', in C. Ward, S. Bochner and A. Furnham (eds) *The Psychology of Culture Shock.* Routledge, New York.

Wellman, B. & Haythornthwaite, C. (eds) (2002) *The Internet in Everyday Life.* Blackwell Publishing, Oxford.

Werbner, P. (1997) 'Introduction: The dialectics of cultural hybridity', in P. Werbner & T. Modood (eds) *Debating Cultural Hybridity.* Zed Books, London, pp. 1–26.

Werbner, P. (1999) 'Global pathways: Working class cosmopolitans and the creation of transnational ethnic worlds', *Social Anthropology,* vol. 7, no. 1, pp. 17–35.

Wilding, R. (2003) 'Romantic love and "getting married": Narratives of the wedding in and out of cinema texts', *Journal of Sociology,* vol. 39, no. 4, pp. 373–389.

Wilding, R. (2006) '"Virtual" intimacies? Families communicating across transnational contexts', *Global Networks,* vol. 6, no. 2, pp. 125–142.

Wilding, R. & Tilbury, F. (2004) *A Changing People: Diverse Contributions to the State of Western Australia.* Department of the Premier and Cabinet, Western Australia, Perth.

Williams, A.M. (1995) 'Capital and the transnationalism of tourism', in A. Montanari and A.M. Williams (eds), *European Tourism: Regions, Spaces and Restructuring.* Chichester, Wiley, pp. 163–176.

Williams, F. (2004) *Rethinking Families*. Calouste Gulbenkian Foundation, London.

Wilson, S.M. & Peterson, L.C. (2002) 'The anthropology of online communities', *Annual Review of Anthropology*, vol. 31, pp. 449–467.

Wimmer, A. & Glick Schiller, N. (2002) 'Methodological nationalism and beyond: Nation-state building, migration and the social sciences', *Global Networks*, vol. 2, no. 4, pp. 301–334.

Wimmer, A. & Glick Schiller, N. (2003) 'Methodological nationalism, the social sciences and the study of migration: An essay in historical epistemology', *The International Migration Review*, vol. 37, no. 3, pp. 576–610.

Wolf, D. (2004) 'Valuing informal elder care', in N. Folbre & M. Bittman (eds) *Family Time: The Social Organization of Care*. Routledge, London, pp. 110–129.

Wood, R. (1980) 'Causes and consequences of increased New Zealand emigration: A commentary', in I. Pool (ed.) *Trans-Tasman Migration*. Population Studies Centre, University of Waikato, Hamilton, pp. 141–150.

Yanagisako, S. (1977) 'Women-centred kin networks and urban bilateral kinship', *American Ethnologist*, vol. 4, no. 2, pp. 207–226.

Yang, B. (1988) 'The relevance of Confucianism today', in J. Cauquelin, P. Lim & B. Mayer-Konig (eds) *Asian Values: An Encounter with Diversity*. Curzon, Richmond, Surrey, pp. 70–95.

Index